# CONSCIENCE IN POLITICS

*Adlai E. Stevenson in the 1950's*

*Men and Movements Series*

# Conscience in Politics

*Adlai E. Stevenson in the 1950's*

◊

STUART GERRY BROWN

Syracuse University Press · 1961

*Library of Congress Catalog Card: 61–13115*

COPYRIGHT © 1961 BY SYRACUSE

UNIVERSITY PRESS

MANUFACTURED IN THE UNITED STATES OF AMERICA

BY THE COLONIAL PRESS, INC., CLINTON, MASSACHUSETTS

# Contents

# Preface

IT IS commonplace to deplore the fact that the American political system has no appointed place for defeated presidential candidates. We give no formal recognition to the leader of the opposition as an officer of government, as do Britain and other parliamentary countries. And our parties refer to defeated candidates as "titular leaders"—with some reluctance and with heavy emphasis on the "titular." It is anything but commonplace, therefore, when a titular leader, as candidate and as opposition spokesman, makes a definable and lasting impact on public policy. Yet the history of the United States from 1952 to 1960 shows that this is precisely what did happen. As candidate and in opposition Adlai E. Stevenson steadily grew in stature as a public man, and his leadership became more rather than less central to the movement of history.

This book is not intended as a chapter in Stevenson's biography, and it is assuredly not conceived as a history of the 1950's. It is an essay in the politics of national leadership —especially leadership by a statesman out of power. It is a study of the remarkable character and effectiveness of Stevenson's public behavior, from the moment of his call to the leadership of the Democratic party in July, 1952, until he relinquished that leadership to John Kennedy in July, 1960.

But the richness of the subject is such that a good many matters—each deserving of special and exhaustive study—are touched upon. In particular, a study of Stevenson's leadership, contrasting inevitably with that of his successful opponent Dwight Eisenhower, raises profound questions about the nature of popularity, of partisanship, and of the relations between both and a man's ability to lead a free people. The evidence here offered suggests that it may be more than mere hypothesis to hold that there is an inverse ratio between wide, unpartisan popularity and leadership; that breadth of popularity may lead to the blurring of issues and the diminution of a man's available power to take the difficult decisions—which are always the divisive decisions. Conversely, the case of Stevenson suggests that on certain transcendent matters, at least, the narrower base of political partisanship may provide effective means for making decisions which answer to the imperative requirements of a historical moment. Thus an opposition leader—even a badly defeated one—may, in the event, provide a leadership the victor cannot exercise. The flaccid body of a democratic nation requires at times the taut toughness of conscience. Such questions go to the heart of presidential leadership in the United States and suggest, in turn, that that much-discussed subject is still far from exhausted.

Stevenson's list of achievements is remarkable on any showing, but especially so since his "control" over his party was firm only during the short intervals of his two campaigns. Most of the time and on most of the issues he was forced to rely on informal advisers and, above all, on his own ability to persuade Americans through the spoken word, without benefit of recognized authority to support him.

Stevenson's first achievement, at least in chronological

sequence, he shares with Dwight Eisenhower. There can be no doubt that the nomination of these two men in 1952 raised the public's expectation of the qualities of mind and character that could be summoned to the Presidency. The public behavior of both men in the ensuing years and their renomination in 1956 confirmed the expectation.

When it came to formulating and acting on issues before the nation and the world, however, Eisenhower and Stevenson immediately took divergent paths. To some extent Eisenhower's lax reins on his administration afforded Stevenson an opportunity for leadership he might not otherwise have had. But profound differences in political philosophy and in assessing the course and meaning of history were of greater importance to Stevenson. He could and did make his own opportunities for leadership, both in suggesting lines of policy and in dealing with political issues.

McCarthyism is a case in point. While Eisenhower felt it necessary to temporize with McCarthy and his followers during the 1952 campaign, and after his assumption of office seemed content, on the whole, to let matters drift, Stevenson did not hesitate to face the issue directly during his campaign and could exert decisive influence afterward.

On civil rights Eisenhower had the initial advantage. His party was not dependent on the South and had a popular issue in the North. Stevenson, for his part, had to "contain" a national party which itself largely "contained" the issue of racial discrimination. Yet it was Eisenhower who allowed too much patience to sap his strength as a leader, while Stevenson took the positive position and showed the way to the nation when the Supreme Court ordered the integration of the public schools.

In foreign affairs Eisenhower's leadership was episodic. Stevenson, on the other hand, from his firm stand on Korea in 1952, through his campaign in 1956, to the dramatic and disheartening events of the spring of 1960 showed steady judgment and creative purpose. While his support of the Eisenhower administration at critical moments was both patriotic and generous, on certain overriding issues he proposed a different course. On such matters as the repeated war scares over Quemoy and Matsu, dealing with the Middle Eastern crisis, and testing H-bombs, Stevenson's opposition views have tended to prevail. When in November, 1957, the Eisenhower administration called urgently upon him for assistance in formulating policy for NATO, the enthusiastic reaction of editors both in the United States and abroad attested to Stevenson's stature as a world statesman.

When the summit meeting of 1960 collapsed in the wake of the U-2 spy plane incident, and rioters in Japan protested renewal of its treaty with the United States, the Eisenhower foreign policy lay in a shambles. The President himself had his invitation to Russia revoked and was advised to cancel his visit to Japan. Simultaneously, editors throughout the free world hinted broadly their hope that the Democrats would defy precedent by nominating Stevenson a third time, and a strong popular movement to draft him—too late to be "practical"—attested to the widespread public recognition of his unique stature.

The following pages may serve to show that, though he was to be denied the Presidency, the times may not, in fact, have been so out of joint for Stevenson as the election tallies suggested.

During the lengthy period of preparation of this book—which dates in part from 1957—I have had the opportunity to try out ideas on a number of people who have been interested in what I was trying to do, critical at times, and always helpful. I should like here to thank the following friends, while absolving them from any share of responsibility: Cass Canfield, Harlan Cleveland, Finla G. Crawford, Thomas K.

Finletter, Roscoe Martin, Edward E. Palmer, Clinton Rossiter, Michael O. Sawyer, Arthur Schlesinger, Jr., T. V. Smith, Sidney C. Sufrin. I am grateful to Miss Carol Evans, Governor Stevenson's long-time personal secretary, for her invaluable assistance. My friend Charner Perry has kindly allowed me to reprint here portions of two articles which first appeared in *Ethics,* of which he is editor (Chapter II on McCarthyism and Chapter III on civil rights). Senator Eugene J. McCarthy made available to me the tape recording of his 1960 speech nominating Stevenson at Los Angeles.

I am grateful, as always, to my wife, for support and useful criticism far beyond the call of duty.

Finally, I should say that Governor Stevenson has responded with characteristic courtesy to my demands upon his time to clarify for me complex matters of historical fact, and has generously made his papers available to me. Needless to say, he has no responsibility whatever for what I have done with them.

STUART GERRY BROWN

*Maxwell Graduate School of Citizenship and Public Affairs*
*Syracuse University*
*June, 1961*

# CONSCIENCE IN POLITICS

*Adlai E. Stevenson in the 1950's*

## CHAPTER I

# Two Candidates:
# Qualities of Leadership

## I

DURING the summer of 1952 an American citizen looking at the political situation in his country had reason to congratulate himself. Whatever his views on the epoch then coming to a close, the future was promising. Though there was a bitter war going on across the world in Korea, there were signs that the stalemate there would end in truce and, though the cost was great, international communism would once more be contained. At home the nation prospered, and there was a choice between two reluctant but remarkably attractive candidates for the Presidency. Whoever won in the November elections, the nation would be refreshed. If it should be the Republicans, under the leadership of Dwight D. Eisenhower, millions of rank-and-file citizens and thousands of politicians would experience relief from a twenty-year political frustration. If the Democrats should win, under the leadership of Adlai E. Stevenson, other millions of rank-and-file citizens would be encouraged to look forward with greater confidence than ever, and thousands of politicians would feel the invigoration of new leadership. For all Americans, in any case, there would be change.

1

The human quality of the candidates was, above all, symbolic of promise for the future. Old timers recalled the 1916 election, when Woodrow Wilson opposed Charles Evans Hughes. One had to go back that far to find a fair comparison. That Eisenhower and Stevenson were obviously men of the finest integrity and dignity, and in their various ways richly experienced in the affairs of the nation and of the world, was the country's good fortune. That they were very different men in manner, outlook, and background—to say nothing of party—was the country's challenge to decision.

Eisenhower's name was already a byword. Reared on a farm in Kansas, he had won admission to West Point and made his career as a soldier. His fine intelligence had improved the years with discipline and training, and he had mastered the arts of war. He was ready when the Second World War called him to greatness. His command of the Allied forces in Europe brought him renown such as few soldiers have known, and assured that he would remain at the center of mighty events. His tenure of the presidency of a great university was shortly interrupted by his call to serve as commander of the new international army established by the Western Allies to hold back the aggressive forces of communism.

As early as 1946 there had been talk of his running for President. In 1948 people in both parties made overtures to him. By 1951 it was clear that he could not escape the growing pressures. He must presently announce both his choice of party and his availability as a candidate. No one can fairly question Eisenhower's sincerity in declining as long as he could to run for President. He entertained profound doubts about the propriety of a professional military man assuming the highest civilian office, and genuine concern as to his own fitness in view of his lifelong separation from domestic politics and issues. But in the spring of 1952 he declared himself a Republican and came home to campaign for the nomination. He challenged and defeated Robert A. Taft, the best-loved and most skillful professional politician in his party.

On July 12, in the Chicago Stockyards amphitheater, Eisenhower accepted the nomination. Addressing not only an enthusiastic convention but also many millions of citizens over radio and television, he took over the leadership of the Republican party and opened his campaign:

> You have summoned me . . . to lead a great crusade . . . for freedom in America and freedom in the world. I know something of the solemn responsibility of leading a crusade. I have led one. I take up this task, therefore, in a spirit of deep obligation. I accept your summons. I will lead this crusade.

Everyone who had followed the General's career knew that it had not been easy for him to reach this point. He would have preferred to stay in military harness, or go back to Columbia University. And there would have been solid satisfaction for him. But now that he had reached the point of no return, the words came with surprising ease. Somehow a political campaign and a political mission could be quickly transformed into a "crusade." Military language could be useful in politics. Forty years earlier an amateur soldier but professional politician had aroused a convention to frenzy when he told them, "We stand at Armageddon and we battle for the Lord!" The poetry of religion somehow infused the military summons of both the Colonel and the General.

How would the General define his crusade?

> Our aims—the aims of this Republican crusade—are clear: to sweep from office an administration which has fastened on every one of us the wastefulness, the arrogance and

corruption in high places, the heavy burdens and the anxieties which are the bitter fruit of a party too long in power.

A Democrat might stumble over the notion that crusades can be either Republican or Democratic. An impartial observer might wonder whether sweeping a party out of office was a suitable objective for such heroic measures. But it was good politics. Partisan politics is seldom fair and often rough. Feeling was high. Republicans did not wish to hear a sweetly reasonable voice. They had nominated a soldier to lead them to victory, and this was the way to do it.

But the soldier had a broader vision. To beat the Democrats was not enough.

> Much more than this, it is our aim to give to our country a program of progressive policies drawn from our finest Republican traditions; to unite us wherever we have been divided; to strengthen freedom wherever among any group it has been weakened; to build a sure foundation for sound prosperity for all here at home and for a just and sure peace throughout our world.

Again, a Democrat might wonder how a partisan crusade would manage to unite a divided people. An outsider might wonder what a "sound prosperity" might be and how one builds a "sure foundation." But such doubts have no real relevance to such a speech under such circumstances. Politics has its rituals. Candidates must be for "unity" and "prosperity" and "sound foundations" at the same time that they must arouse their followers and persuade still others to follow them. What was remarkable about the General's address was not that the ritual of his speech, like any other, was open to critical question, but that what he said was indeed ritual.

It did not seem so at the time. The exciting atmosphere of the convention, the dignity of the speaker, and the heroic aura with which the American people had surrounded him, engulfed what he said. He spoke well and he spoke sincerely. Expectation, perhaps, supplied its own answers to the questions posed by what the candidate did not say.

Yet in the perspective of years Eisenhower's acceptance speech takes on greater interest. It was a true indication of the man, both of his strengths and his weaknesses. In the years of his Presidency there was more, not less, of the grand generality. There was more, not less, of the heroic aura. And there was less, not more, effective leadership. Eisenhower's experience as a military man would prove to be too limited for the Presidency. He had learned how to carry out policy; as President he would be called upon to make it.

But all this was scarcely predictable in July of 1952. In the inner circle of the Republican party Eisenhower's victory over Senator Taft meant control of the party by the same forces—the liberal wing—that had nominated Wendell Willkie in 1940 and Thomas E. Dewey in 1944 and 1948, only to lose their control with loss of the election. While there was now some grumbling among the Old Guard that "organization people" were loyal to Taft as a "true" Republican and would not work for Eisenhower, the overwhelming sentiment was for the General, and sure that *this* time the party would win. Conservatives immediately began their strenuous and continuous "battle for the mind of the candidate." No Republican politician doubted that Eisenhower was henceforth the head of the party.

Nor did the public. Americans watching the Republican Convention on their television screens had witnessed a bitter political struggle between the supporters of Eisenhower and those of Taft. The skillful men in the General's camp, led by Dewey, had scored decisively, not only in putting their candidate over but in making it appear that Eisenhower's victory over Taft was the triumph of good over evil. But when the Senator grasped the General's hand before the television camera, the symbol of evil seemed to disappear in the warmth of the loser's smile. The Senator was "a good soldier" after all. He would await the orders of his commander and serve the "crusade" with devotion.

Richard Nixon had a consistent Old Guard voting record in the House and in the Senate, he had helped to put Alger Hiss in jail, and he had youth. Thus he was politically qualified to join battle on the side of the General. But his nomination for the Vice-Presidency seemed of little consequence at the time. The "crusade" was about to begin. The point now was, who would be chosen to lead the enemy?

## II

Nine days after the General accepted the Republican nomination and announced his crusade, the Democratic Convention opened in the same Chicago Stockyards with a welcoming address by the Governor of Illinois. His appearance on Monday morning, July 21, was the first view of Adlai Stevenson for most Americans, and the words he then spoke the first they had heard. People outside of Illinois knew his name because it had been in the headlines for months. Popular magazines had featured stories of his distinguished family background, including a namesake grandfather who had been Vice President under Cleveland; his education at Princeton, Harvard, and Northwestern; his government service in the Navy Department and in foreign affairs before his election as Governor of Illinois in 1948. There had been continuous speculation over whether he would or would not run for the nomination—would or would not accept the nomination if drafted. Democrats and Republicans alike were, at the least, curious to see and hear this man who had been the object of a national guessing game. What was he like? Was it simply that the Democrats were so torn by faction and dissension that the leaders turned to Stevenson precisely because he was not well known, not a controversial figure—yet governor of a great state? Or was it because those who urged his nomination were aware that he possessed unusual qualities?

His opening was predictable enough. His assignment was to welcome the Democrats to his state, and he welcomed them. His words were well chosen, but there was nothing striking about their meaning. Fairly soon he made a wisecrack. After twitting the Republicans for battling in the stockyards and blasting their Democratic rivals for all kinds of sins, the Governor observed that he "was even surprised the next morning when the mail was delivered on time!" But it was only a light touch in a speech you could expect to have light touches. Then another touch of humor—"perhaps the proximity of the stockyards accounts for the carnage" (the Republican internecine warfare).

But with the next few words the mood and tempo changed sharply:

> The constructive spirit of the great Democratic decades must not die here on its twentieth anniversary in destructive indignity and disorder. And I hope and pray, as you all do, that we can conduct our deliberations with a businesslike precision and a dignity befitting our responsibility, and the solemnity of the hour of history in which we meet.

It was the beginning of a different kind of political oration; and the introduction, as it turned out, of a new eloquence. The point was to turn the political quarrels of the Republicans into a moral for Democrats, and the Democratic party into a moral agent:

> For it is a very solemn hour indeed, freighted with the hopes and fears of millions of mankind who see in us, the Democratic Party, sober understanding of the breadth and depth of the revolutionary currents of the world. Here and abroad they see in us awareness that there is no turning back, that, as Justice Holmes said, "We must sail sometimes with the wind, sometimes against it; but we must sail and not drift or lie at anchor." They see in us, the Democratic Party that has steered this country through a storm of spears for twenty years, an understanding of a world in the torment of transition from an age that has died to an age struggling to be born. They see in us relentless determination to stand fast against the barbarian at the gate, to cultivate allies with a decent respect for the opinion of others, to patiently explore every misty path to peace and security which is the only certainty of lower taxes and a better life.

Republicans, listening and watching, might be pardoned for doubting that the people of the world really "saw" as much as this in the Democratic party, or that Democrats had always acted with such heroic courage or always borne adversity with such heroic patience. Detached observers might consider that virtues tend to be distributed rather more evenly among people and parties, and might fairly ask whether the barbarian would surely have triumphed had someone other than Democrats barred the gate. But the "storm of spears" was not ordinary political oratory, nor were the echoes of Matthew Arnold and Jefferson. And whether or not one was persuaded by what he was saying, one could not doubt the speaker's conviction.

Another change of mood and tempo. Eloquent praise gave way to moral thrust. Republicans were excoriated for superficiality; but so, by implication, were Democrats:

This is not the time for superficial solutions and endless elocution, for frantic boast and foolish word. For words are not deeds and there are no cheap and painless solutions to war, hunger, ignorance, fear and to the new imperialism of Soviet Russia. Intemperate criticism is not a policy for the nation; denunciation is not a program for our salvation. Words calculated to catch everyone may catch no one. And I hope we can profit from Republican mistakes not just for our partisan benefit, but for the benefit of all of us, Republicans and Democrats alike.

This was not convention ritual. It was not the calculating flattery one might expect from a candidate, reluctant or anxious. It was, perhaps, shocking language from a host. Yet by the end of the summer Americans, whether they approved or not, would know that these words were, nevertheless, a true measure of Stevenson's mind and character.

But there was more:

Where we have erred, let there be no denial; where we have wronged the public trust, let there be no excuses. Self-criticism is the secret weapon of democracy, and candor and confession are good for the political soul.

Here was a strange new note to sound at a political convention. Parties just do not criticize themselves or admit weaknesses in public. Such talk might make its way into a book of quotations (as it did) but it could not be good politics. The audience was quietly astonished.

Still, the Governor was a politician. With another quick transition he had his listeners on their feet and cheering:

> But we will never appease, nor will we apologize for our leadership in the great events of this critical century from Woodrow Wilson to Harry Truman! Rather will we glory in these imperishable pages of our country's chronicle. But a great record of past achievement is not enough. There can be no complacency, perhaps for years to come. We dare not just look back to great yesterdays. We must look forward to great tomorrows.

Even the tough-minded members of the working press were stirred. Here was exciting copy.

But would there be a clue to the Governor's tactics? Or were they indeed tactics? Was he perhaps taking advantage of his position of "no personal axe to grind"? Was he moving secretly toward his own nomination? A clue did seem to come presently.

> What counts now is not just what we are *against,* but what we are *for. Who* leads us is less important that *what* leads us—what convictions, what courage, what faith—win or lose. A man doesn't save a century, or a civilization, but a militant party wedded to a principle can.

Was Stevenson saying, "Stop thinking of me as a candidate, and let the best man among the others win"? No one could be sure. Yet, no one could doubt that he was transporting his audience. At the end he returned to the counterpoint of moral exhortation and eloquence, and somehow got away with it:

> So let us give them a demonstration of democracy in action at its best—our manners good, our proceedings

orderly and dignified. And, above all, let us make our decisions openly, fairly, not by the processes of synthetic excitement or mass hysteria, but, as these solemn times demand, by earnest thought and prayerful deliberation.

Thus can the people's party reassure the people and vindicate and strengthen the forces of democracy throughout the world.

Stevenson's speech had actually been a very short one. But it would be hard to find a "welcoming address" remotely like it in the annals of American party conventions. Stevenson had not only shocked, puzzled, annoyed, and aroused the convention to high enthusiasm almost simultaneously. He had also introduced himself unforgettably to his party and to the country. Henceforth no one who had heard him would fail to recognize his voice, or the quality of the man in the words he spoke. That he would be nominated was now certain. But how would he answer to the responsibilities of leadership?

When Adlai Stevenson was "summoned," like Eisenhower, to lead his party, it was in the smaller hours of the morning. The delegates were exhausted from a week of constant and feverish activity—their manners not always good, their decisions not always openly arrived at. And the candidate, unlike Eisenhower, was not, even in the moment of his glory, the sole focus of the convention. A Democratic President was in office, and he spoke first. Harry Truman was a beloved party man. He was entitled to share the occasion, and he did so with his typical gusto. He "gave 'em hell." Stevenson, nominated under protest and accepting with misgiving, could not lead a crusade had he wished to. He faced instead the nearly impossible task of asserting his unwilling leadership of a party loyal, for the most part, to a President from whom he felt a need to achieve independence, and to prove his independence to the nation. It was as complex a political situation as one could conjure. Yet it perfectly suited the complexity of Stevenson himself. He could and did reveal certain aspects of his intellect and character to best advantage under such circumstances.

From the outset the contrast with Eisenhower was clear and dramatic:

> I accept your nomination—and your program.
> I should have preferred to hear those words uttered by a stronger, a wiser, a better man than myself.

Later, after the November votes were counted, it would be suggested that Stevenson lost the election when he was overheard to tell the Illinois delegation, before the convention, that he thought himself unfit for the Presidency. It might well be said that these opening words of his acceptance address were sufficient confirmation for such a view. But the election was months away, and a discerning listener could find something more than self-depreciation as Stevenson elaborated his theme. His humility was not the humility of ritual; it was uttered from the depths of a mind which had deeply searched itself. Not only *could* he not seek the nomination he had won because he was already committed to trying for a second term as Governor of Illinois, but

> I *would* not seek your nomination for the Presidency because the burdens of that office stagger the imagination. Its potential for good or evil now and in the years of our lives smothers exultation and converts vanity to prayer.

If some of the voters saw weakness in such sentiments, others perceived in them a second level of meaning. Beyond humility there was a calculated contrast with the positive note Eisenhower had struck. By exalting the office of the Presidency, Stevenson perhaps hoped to make the General appear overconfident, and to prepare the way for an appeal to his party to help him, to make the campaign a common enterprise—to make the party in some sense responsible for victory or defeat:

> And now, my friends, that you have made your decision, I will fight to win that office with all my heart and soul. And, with your help, I have no doubt that we will win.

The last bit was surely rhetoric. The whole tone of his address showed how much he doubted. But since Eisenhower had accepted a "summons" to lead and had made it clear to the Republicans that this was exactly what he intended to do, Stevenson threw as much of his burden as he could upon others, both because he desperately felt the need of their help and because such a posture would rescue him from the hopeless task of trying to match Eisenhower's renown as a leader. He would come back, in the conclusion of his address, to this major theme.

If he was to be a leader among equals—not a commander—he would find it necessary to contrast his cause dramatically with the crusading procession of troops behind his opponent. He succeeded well enough in making the contrast, though events proved that the dramatic aspect left something to be desired:

> What does concern me, in common with thinking partisans of both parties, is not just winning the election, but how it is won, how well we can take advantage of this great quadrennial opportunity to debate issues sensibly and soberly. I hope and pray that we Democrats, win or lose, can campaign not as a crusade to exterminate the opposing

party, as our opponents seem to prefer, but as a great
opportunity to educate and elevate a people whose destiny
is leadership, not alone of a rich and prosperous, contented country as in the past, but of a world in ferment.

"Thinking" was to oppose emotion; education would
counter the crusading spirit. Here was the theme of the campaign, whether fighting Harry Truman liked it or not. The
latter's misgivings, which for a time in 1956 were to become
irrepressible doubts, may well have commenced at this moment. Yet who can say that Stevenson was wrong? What
better way was actually available than to counter crusading
ardor with an appeal to "thinking" people? Assuredly it did
not succeed; but as Americans came to know Stevenson they
could at least agree that the nature of the man made any
other strategy impossible.

As he progressed, the theme of reason and responsibility
took on larger proportions:

> . . . More important than winning the election is gov-
> erning the nation. That is the test of a political party—
> the acid, final test. When the tumult and the shouting die,
> when the bands are gone and the lights are dimmed, there
> is the stark reality of responsibility in an hour of history
> haunted with those gaunt, grim specters of strife, dis-
> sension and materialism at home, and ruthless, inscrutable
> and hostile power abroad.

Here was an echo of the self-criticism theme of his welcoming
address, and, Republicans could quickly add, a flat contradiction of his assertion only a few sentences earlier that Americans were "contented." But if he could not entirely escape
obeisance to political ritual, he could and did quickly leave
it behind:

The ordeal of the twentieth century—the bloodiest, most turbulent of the Christian age—is far from over. Sacrifice, patience, understanding and implacable purpose may be our lot for years to come. Let's face it. Let's talk sense to the American people. Let's tell them the truth, that there are no gains without pains, that we are now on the eve of great decisions, not easy decisions, like resistance when you're attacked, but a long, patient, costly struggle which alone can assure triumph over the great enemies of man—war, poverty and tyranny—and the assaults upon human dignity which are the most grievous consequences of each.

The night was very late. Delegates were stiff with fatigue, and viewers and listeners still following in their homes were scarcely fresher. And the sentences were complex. One could fairly wonder whether they made much impression at the time. Yet here was the special seal of Stevenson, not only in the campaign that followed but in all the years thereafter. To talk sense—to tell the truth—was proposed for those who wished to follow him as the better alternative to slogans, generalities, and complacency. In retrospect it might seem almost as though he were preparing to lead an opposition rather than to be President. In his secret heart it may have been so. The voters determined the outcome for him, in any case, and he was indeed prepared to make the most of strategy and circumstance in the years after his defeat.

But the man of reason had a rapier in his hand. The victory to be won over the enemies of man, he said, "mocks pretensions of individual acumen and ingenuity."

For it is a citadel guarded by thick walls of ignorance and of mistrust which do not fall before the trumpets' blast or the politicians' imprecations or even a general's baton.

In the end he came back to his need for help, to share his
burden with his party, and to the note of self-effacing humility which opponents would ridicule and friends would
respect:

> That, I think, is our ancient mission. Where we have de-
> serted it we have failed. With your help there will be no
> desertion now. Better we lose the election than mislead the
> people; and better we lose than misgovern the people.
> Help me to do the job in this autumn of conflict and cam-
> paign; help me to do the job in these years of darkness,
> doubt and of crisis which stretch beyond the horizon of
> tonight's happy vision, and we will justify our glorious
> past and the loyalty of silent millions who look to us for
> compassion, for understanding and for honest purpose.
> Thus we will serve our great tradition greatly. . . . And
> finally, my friends, in the staggering task you have assigned
> me, I shall always try "to do justly and to love mercy and
> to walk humbly with my God."

Cynics would call the whole speech mawkish. Republicans
would assail Stevenson as a weakling. Dispassionate analysis,
of which there was little in the summer of 1952, could find,
after the paradox of the welcoming address, a revelation of
style and character. Democrats, in the first glow of discover-
ing a new leader, talked of a "prairie Roosevelt" or a "new
Wilson." Yet, the event disclosed Stevenson as a singular
kind of politician with a fresh approach to politics and affairs
of state. The circumstances of his nomination in some measure
dictated both the strategy of his acceptance and the tactics of
his campaign. They could not reveal the whole man. It took
four years in opposition and a second try for the Presidency
to approach that. But, "win or lose," as he himself so often
said, the very power of his eloquence moved him onto the

public stage and decreed that he should stay there, at or near the center.

Here, then, were the two candidates before the public in the summer of 1952: both new to the national political scene; both reluctant; both yielding to very great pressure; both approaching the Presidency with high seriousness; each offering the country a significant change. But comparisons quickly ran out. Eisenhower, overcoming his initial diffidence, was bold and commanding. He painted the picture of the present in terms so dreadful that nothing less than a "crusade" would save the country. He was prepared to lead it. His campaign would draw on the symbols of military victory and the church militant. Stevenson, accepting his assignment but forced by his nature to live with his doubts, offered reason, self-criticism, and responsibility as his alternatives. He proposed not to command but to share his leadership. He would draw upon the poetic imagination for his symbols—and on the religious tradition of the lonely search for salvation. Two kinds of vision; two forms of wisdom; two versions of courage; contrasting conceptions of leadership. The soldier or the civilian? The commander or the strange politician? Which would be elected? Which would turn out to be the statesman? These, or something like them, were the questions Americans faced in 1952. But underneath the surface another question was stirring. Which America would the people choose—the America of Eisenhower or the America of Stevenson? And did Americans really know which America they were choosing by their votes?

Given the men and the issue between them it is not surprising that the election produced the largest vote in the history of the country. Though their shares were strikingly unequal, Eisenhower's winning total, greater than any other winner's, was followed by a total for Stevenson greater than any other loser's. More people were stirred by the men, the campaign, and the times, and fewer were indifferent than ever before. But Eisenhower's victory was decisive. How did it come about? What did it mean?

## III

Skillful techniques and impartial evaluation have brought the measurement of public opinion into the realm of reliable prediction. Scientific sampling of opinion during election campaigns was introduced almost unnoticed at the very moment when old-fashioned "straw polls" disappeared with the *Literary Digest* in 1936. Steady improvement in constructing questions, in allocating samples, and in statistical methods had reached a point by 1952 when the outcome of a presidential election could be predicted with a good deal of confidence. Even the mistakes made by the pollsters in 1948 were traceable more to misinterpretation of accurate data than to faulty techniques. In 1952 the polls showed General Eisenhower ahead from the beginning, and though the gap between him and Stevenson appeared to narrow as the weeks went by, no counter tide ever actually developed. The "normal error" of the sampler's poll turned out to favor Eisenhower, so that he won by two or three points more than was predicted. But the polls were substantially correct, as they were again in 1956.

In recent years the experts have been trying to make better analytical use of their data and to sound opinion to greater depths. Thus it was predicted in 1952 that a majority of women would have a decisive influence upon Eisenhower's victory, and follow-up studies showed that the prediction was correct. Similar techniques produced evidence that the Korean War was the most decisive single issue. Corruption, communism, and other issues were important but secondary.

Yet to say *what* people were thinking and *how* they would vote was far simpler than to say *why*. At bottom our understanding of popular feeling and expression still depends upon subjective judgments. No accounting for the Eisenhower victory of 1952 can avoid the use of impressions.

The Korean War is a fair, and crucial, example of the difficulty. There can be no doubt that the war was deeply unpopular. It was being fought far across the world in an area where Americans could feel only the remotest sense of involvement. It was supposed to be a United Nations "police action," yet most of the fighting and much of the dying was being done by Americans. Americans, whose sons and husbands and brothers had only just begun to settle down after the horror and the dislocations of the Second World War, saw in Korea a symbol of the destruction of their hopes for peace and privacy. The war itself was not like other wars. President Truman, the leaders of the government, and the United Nations itself insisted on limited action, while a great soldier like General Douglas MacArthur had wished to launch an all-out attack on the enemy. Stalemate and frustration seemed to be proposed as a policy to a people whose only experience of war was winning.

Yet there was another side to the story. When, in the years after 1945, the Soviet Union broke one by one its wartime agreements with the Western Allies, Americans came to hate and fear communism as never before. They had learned to accept the need for a peacetime standing army, navy, and air force and had supported dramatic and expensive policies directed against communism, like the Truman Doctrine of defending Greece and Turkey, the Marshall Plan for European reconstruction, and the North Atlantic Alliance. In 1948 they endorsed the leadership of President Truman, and in 1950, after the outbreak of the war in Korea, they again

gave him a Democratic Congress. It was the Communists—
Korean, Chinese, and Russian—who had commenced aggression. It was President Truman who had rallied the free world
to stop it. The war had in it the elements of a "crusade"—
the crusade of freedom against Communist tyranny. Why,
then, had the demand to be done with it become so widespread by the fall of 1952? There is no inclusive answer.
But it is certain that the Korean "crusade" was bogged down
in frustration everywhere—among the soldiers at the front,
the administration and Congress in Washington, the diplomats
at the United Nations, and the people in their homes. The
whole world was tired. Americans had had enough of war
and crisis, and no end was in sight. Stalin was too remote to
be punished or to serve as a useful scapegoat. *Someone* had
to be blamed because *everyone* was to blame.

Eisenhower and Stevenson reacted very differently to the
Korean crisis, perhaps inevitably. Stevenson could not repudiate the policies of President Truman had he wished to,
which he did not. Believing that the cause was right and that
Truman's leadership had been both correct and courageous,
Stevenson pledged himself to carry on the war until armistice should mean the achievement of the aims of the United
Nations. In his speeches he sought words to explain the war
to the people and to encourage them to tighten their belts
and "take it" so long as might be necessary:

> How long can we keep up the fight against the monster
> of tyranny? How long can we keep on fighting in Korea;
> paying high taxes; helping others to help ourselves? There
> is only one answer. We can keep it up as long as we have
> to—and we will.
>
> That is why we cannot lose, and will pass from dark-
> ness to the dawn of a brighter day than even this thrice-
> blessed land of ours has ever known.

While many Americans might assent to such exhortations, it was impossible to be enthusiastic. It can never be popular politics to tell people, however eloquently, that they must continue indefinitely to do profoundly unpleasant things.

If dealing with the Korean problem was an impossible political assignment for Stevenson, it was scarcely less difficult for Eisenhower. As commander of the NATO forces, and earlier as Chief of Staff of the Army, Eisenhower had played an important role in the American and Allied effort to contain communism and resist aggression. He could not have been, and was not, opposed to the Korean War. His was a genuine dilemma: either accept Stevenson's proposal that the Korean War be eliminated as a campaign issue, which might be an act of political suicide, or address it critically as the prime issue. He chose the latter course. That it was the easier course no one could question. And it may be that by thus making the popular decision at the outset of his political career he both assured his popularity and sapped his moral strength for the years ahead. Once he had begun to tell the people what so many wanted to hear—that our plight was due to "bungling," that the boys should be brought home ("Let Asians fight Asians"), and finally "I shall go to Korea"—once he had moved in this direction he might never again be able to check a popular tide, even when he deplored its course. He might find himself, instead, frequently drifting on the waves of his own popularity.

As politics of the moment, however, it was an astute choice. Seldom has the image of a military hero so perfectly answered to the yearnings of a people. The leader of the "Crusade in Europe" now assumed the role of leader in the most popular of all crusades—the crusade for peace. If his spirit was troubled by a sense of shifting from one set of convictions to another under force of circumstance, he could, and apparently did, identify his new turn with his well-earned hatred for bloodshed and his profound hope for peace. To be known in history as a peacemaker became the measure of his aspiration. Thus he built a bond between himself and the people that would withstand almost any attack, even, as it turned out, the repeated attacks of illness. The bond he forged with the people was not the partisan bond that ties a devoted political leader to a devoted political party, like Franklin Roosevelt to the Democrats or Robert A. Taft to the Republicans. Partisans must make enemies simply because they are partisans. Peacemakers cannot afford to make enemies. As the years went by, Eisenhower left his party ever further behind him, and rose ever higher in popularity. He practiced the art of avoiding commitment on sharp issues with undiminishing success, until a major economic recession and the dramatic appearance of Russian sputniks began to erode the popular confidence in him. But in the campaign of 1952 his transition from active leader to symbol was only beginning.

At the outset Eisenhower made a skillful ascent toward the plane above partisanship in domestic matters. While he rejected Stevenson's proposal to eliminate Korea from the campaign, he forced Stevenson to share with him the strongest planks of the Democratic platform—the successful reforms and programs of the Roosevelt administration. These reforms, he said, were no longer at issue. Social security, collective bargaining by strong labor unions, minimum wages and maximum hours, and federally supported prices for agriculture were some of the Democratic achievements that would henceforth lie beyond politics. Stevenson could scoff and question; he could point to the long record of Republican opposition in Congress, and he could score with irony and humor. But many of his weapons were nevertheless blunted.

On corruption, domestic communism, high living costs,

and other lesser issues Eisenhower played well his unaccustomed part of finger-pointing scourge. The times were friendly. The issues were partisan, but even many Democrats were disturbed by the parade of officials in the Truman administration who had been exposed in graft and improper influence, by the growing list of government employees accused by Senator McCarthy as friendly to communism and the Soviet Union, and by the inflation of prices they had to pay. In the great halls at mass rallies and on the rear platform of his train General Eisenhower became genial "Ike," yet somehow remained the exalted and revered hero-General. In nearly the same breath he could both inspire confidence that he would restore order to dishonorable chaos at home, and seem to promise peace and an end to Communist aggression. Republicans were elated, and countless others decided it was "time for a change."

Meanwhile, Stevenson fought his losing fight with growing respect from a people who had already decided in their majority against him. After his defeat he received an unprecedented avalanche of letters from people who had voted for Eisenhower with a somewhat uneasy conscience. They admired Stevenson and believed him when he spoke. But they thought that the odds were against him. As a Democrat it would be harder for him to "clean up the mess in Washington"—though he had naively assured one newspaper editor that he could, while President Truman fumed that there was no mess to clean up. Stevenson, the letters said, was no less opposed to communism in government than Eisenhower, but he would inherit an administrative apparatus that he could not make his own. And he could not, as Truman's successor, make a fresh start on Korea, or so they thought. These letters to the defeated candidate from those who had defeated him seemed a fair symptom of popular feeling. If the odds he faced in the campaign were nearly hopeless, that he received so tremendous a losing vote was an unprecedented tribute.

Korea hung in heavy clouds over the whole campaign. It was Eisenhower, not Stevenson, who seemed to point out the slanting rays of sunlight—rays which could be broadened to the clear light of day if he were elected. In retrospect, Korea seems to have made the crucial difference—Korea and the General. Parties charged with corruption had been victorious before. There was no precedent for a party's being turned out in prosperous times. Even national heroes had been defeated for the Presidency—like General Fremont and Theodore Roosevelt. But when the voters went to the polls in November, 1952, the usual currents were not running. What was left of old loyalties among the voters was reflected in the hairline margins of the congressional elections. What was new was revealed in the mighty wave of sentiment for Eisenhower.

Stevenson himself afterward set down as good a brief accounting as any:

> . . . as the battle of words progressed, I felt more and more that people cared little about the issues and party records, or about precise definition of positions. They were weary of conflict, impatient and eager for repose.

Stevenson denied that "repose" was in sight. Eisenhower seemed to point to it—something you could have just after the election.

## IV

At the moment of defeat Stevenson added another dimension to his stature before the country and the world. He could turn the loser's ritual of concession into a sign of victory for all lovers of freedom:

My fellow citizens have made their choice and have selected General Eisenhower and the Republican Party as the instruments of their will for the next four years.

The people have rendered their verdict and I gladly accept it. General Eisenhower has been a great leader in war. He has been a vigorous and valiant opponent in the campaign. These qualities will now be dedicated to leading us all through the next four years. It is traditionally American to fight hard before an election. It is equally traditional to close ranks as soon as the people have spoken. From the depths of my heart I thank all of my party and all of those independents and Republicans who supported Senator Sparkman and me.

Felicity of phrase and generous good sportsmanship thus far. It would have been enough simply to finish by reading a telegram of congratulation to Eisenhower. But there was more, and the more was what mattered:

That which unites us as American citizens is far greater than that which divides us as political parties. I urge you all to give General Eisenhower the support he will need to carry out the great tasks that lie before him.

I pledge him mine.

We vote as many, but we pray as one. With a united people, with faith in democracy, with common concern for others less fortunate around the globe, we shall move forward with God's guidance toward the time when His children shall grow in freedom and dignity in a world at peace.

Then the telegram:

The people have made their choice and I congratulate you. That you may be the servant and guardian of peace and

make the vale of trouble a door of hope is my earnest
prayer.

The losing race was over for the troubled, soul-searching
politician. His campaign of exhortation to endless effort, patience, and acceptance of heavy responsibility, with a better
world only far down a "misty path," had ended in decisive
defeat. But those who saw in him at the outset a Hamlet
complex had long since been proved mistaken. His campaign had shown the people a man whose soul-searching led
to decision, whose patience prepared the way for action.
And they had learned to expect humor. Stevenson's sense for
the tragic was balanced by a sense for the comic that preserved him from arrogance. At the very end he remembered
an old story. It showed his opponents that he could "take it";
it relieved the feelings of his followers and, perhaps, himself:

> Someone asked me, as I came in, down on the street, how
> I felt, and I was reminded of a story that a fellow-towns-
> man of ours used to tell—Abraham Lincoln. He said he
> felt like a little boy who had stubbed his toe in the dark.
> He said that he was too old to cry, but it hurt too much to
> laugh.

General Eisenhower, in his moment of triumph, had not
only the good wishes of his opponent but the immeasurable
enthusiasm of his supporters. He acknowledged his applause
with his well-loved grin and his newly patented victory sign
—both arms stretched high above his head and outward for
the letter "V." In the weeks of the campaign he had never
mentioned Stevenson by name, for good political reasons.
Whatever the reasons, he did not mention the name now.
His telegram to the Governor was polite but formal:

I thank you for your courteous and generous message. Recognizing the intensity of the difficulties that lie ahead, it is clearly necessary that men and women of good will of both parties forget the political strife through which we have passed and devote themselves to the single purpose of a better future. This I believe they will do.

His words to his followers and to the country were words appropriate to a successful "crusade." But it was a personal victory, his and the people's, not the victory of ideas or definable goals:

. . . I am indeed as humble as I am proud of the decision that the American people have made. And I recognize clearly the weight of the responsibility you have placed upon me, and I assure you that I shall never in my service in Washington give short weight to those responsibilities. To each of you who has worked so hard to make this crusade a success so far, to every man, woman and child —I extend my warm thanks and hopes that the day will come when I can extend that thanks in a more personal way. . . . And I also point out that we cannot now do all the job ahead of us except as a united people. And so let us really put into practice what I have tried to do so haltingly in the little telegram that I sent to my late rival.

Thus was the image of the new President brought into being. Eisenhower had climbed above issues to a peak of nonpartisan popularity that few Americans have reached. To stay there he would have to be permanently disengaged from great battles. He had imposed upon himself, and the people had imposed upon him, conditions that made leadership possible only when his followers were nearly unanimous. But the

issues among free peoples are partisan issues. On such matters as McCarthyism and civil rights, the most compelling
domestic issues of his Presidency, Eisenhower was remote
and ineffectual. Decisions, of course, were made. But someone else, someone in the administration like Richard Nixon, or
Herbert Brownell, or Sherman Adams, or George Humphrey,
or above all, John Foster Dulles, always seemed to make
them. If there was credit, the people gave it to the President. If there was blame, the people seemed to attribute it
to someone else. In foreign policy, when it was a matter of
universal ideals, Eisenhower was a fitting symbol, as at
Geneva in 1955, or when he pleaded for the peaceful use
of atomic energy before the United Nations; but when it was
a matter of action, when opinion was divided, his very popularity often seemed to doom him to vacillation and weakness.

But Stevenson, the partisan politician, was free to criticize
and advocate, to plead and exhort and berate until decisions were taken, when both the rules of democracy and
his own temperament required him to support the government. His freedom as a partisan enabled him to be decisive
where Eisenhower was hesitant, to influence where Eisenhower was aloof, to propose new ideas and new policies, and
to lead by articulating for others when a majority was forming.

In the campaign of 1952, American citizens knew that
they were choosing between Eisenhower and Stevenson; they
may not have realized that they were choosing both. The
images of both men were to some extent distorted by the
conditions of the campaign itself and the political background
on which it was staged. The partisan side of Eisenhower,
underlined during the campaign by his attacks on corruption,
communism, and bungling, soon receded and reappeared
over the years only spasmodically. The reluctant, self-depre-
ciating side of Stevenson, so marked at the moment of his

nomination, resumed its proper and smaller proportion when defeat relieved him of responsibility for defending another man's record. The years after November 5, 1952, corrected the perspective on both men. They were the signs of two different Americas—the one a relaxed and disengaged America the people yearn for and twice voted for at the polls, the other an anxious but committed America the times still imperatively call for.

## CHAPTER II

# Stevenson vs. Eisenhower
# in the McCarthy Era

I

JOSEPH R. McCARTHY, junior senator from Wisconsin, made his sensational entrance on the public stage at Wheeling, West Virginia, on the evening of February 9, 1950. The era which bears his name had a prologue, perhaps in the days of the Dies Committee just before the Second World War, but its ending coincided remarkably with McCarthy's fall. He began with an attack on the Department of State and its Secretary, Dean Acheson, for maintaining in their employ persons "known" to be Communists or agents of international communism. He assumed the role of public prosecutor, and charges and counter charges became the hallmark of his arrogant presumption to replace the courts of law. The term "McCarthyism" quickly came into being. To the Senator's friends it meant a heroic struggle to rid the government of subversives and cleanse the nation of spies and traitors and their dupes by "exposing" them to public censure. To his opponents it meant a calculating technique of smear and slander that imperiled the civil liberties at home and demeaned the image of America abroad.

McCarthy's first speech at Wheeling foreshadowed all

the rest. He charged that there were 205 members of the Communist party in the State Department. Or was it 57? Or was it 81? The charge stuck in the public mind, while the discrepancies about numbers served only to keep the matter in the headlines. Whatever may have been his purposes, the headlines were certainly his means. And he kept himself in them for more than four years, often overshadowing a war and two Presidents. McCarthy made news, while others nervously waited for him to make it.

Before 1953, McCarthy's attack was aimed at Democrats. Many Republicans found him useful in their campaigns to "turn the rascals out." Sometimes Republican leaders and candidates felt it necessary to repudiate specific charges McCarthy made, but they could and did profit from the suspicion he was creating that "where there is smoke there is fire." By the summer of 1952 his charges had been investigated by one committee of the Senate and he himself had been investigated by another. His reply to the findings of the Tydings Committee, which dismissed his charges as unsubstantiated, had been to intervene in the Maryland elections of 1950 and help by dubious means to defeat Senator Millard Tydings. He had branded Senator William Benton of Connecticut, and others who sought to expose the emptiness of his accusations, with the repeated charge that they were themselves the dupes of communism. His charges of subversion and disloyalty undercoated his attacks on Democratic foreign policy in the Far East as a disaster which had lost China to the free world. He had even reached out to impugn the loyalty of General George C. Marshall, the wartime Chief of Staff who had served as Secretary of State and the Secretary of Defense in the Truman administration.

McCarthy was a Republican and was counted among the stalwart members of the Old Guard. But his charges were persuasive, too, among Democrats, especially among conservative Democrats who had been uneasy during the reforms of the New Deal and considered President Truman too "liberal" or too "pro-labor." He aroused old fears of aliens and played upon the newer fears of Communists and Russians. Not only must a Democratic candidate for President prepare to defend himself and his party against McCarthy's onslaught; a Republican candidate, too, would have to take McCarthy's measure. To millions of Americans one of the most interesting questions in the summer of 1952 was how General Eisenhower would deal with the "Great Accuser."

## II

The first suggestion of the General's tactics came on August 2, when he called for support of all Republicans. "I want Republicans elected so they can organize the Senate and House." But, with McCarthy perhaps implied, he went on, " . . . This does not mean that I have to endorse every single idea of every candidate." The difficulty with this position was that McCarthy was firmly identified in the public mind with one idea only—attacking and exposing "Communists" by his arrogant method of public denunciation. Eisenhower had thus somehow to disengage his support of McCarthy from support of McCarthy's "idea." One of the General's most important backers soon, perhaps unwittingly, forced the issue.

Paul G. Hoffman, former administrator of the Economic Co-operation Administration (Marshall Plan), automobile executive, and at that time president of the Ford Foundation, testified on behalf of Senator Benton, who was being sued by McCarthy for libel and slander. Mr. Hoffman cited, as striking examples of McCarthy's methods, the charges made by McCarthy a year earlier against General George C. Marshall. While it was Senator William Jenner of Indiana, an avowed admirer of McCarthy, who specifically called Marshall a "front man for traitors" and a "living lie," McCarthy, on June 14, 1951, delivered a 60,000-word speech on the floor of the Senate intended to show that the success of the Chinese Communist revolution and other Communist successes were the result of a conspiracy by high officials of the American government, notably General Marshall. The speech was afterward published as a book, but remained privileged from libel actions because it was entirely quoted from the *Congressional Record*. In the course of the speech McCarthy asserted that Marshall was "steeped in falsehood," and that he had "recourse to the lie whenever it suits his convenience." He "reviewed" Marshall's career, as Chief of Staff during World War II, as special emissary to China, and as Secretary of State, in a way to suggest that every success of the Soviet Union or of other Communists was the product of Marshall's treachery:

> This must be the product of a great conspiracy, a conspiracy on a scale so immense as to dwarf any previous such venture in the history of man. A conspiracy of infamy so black that, when it is finally exposed, its principals shall be forever deserving of the maledictions of all honest men. . . . What can be made of this unbroken series of decisions and acts contributing to the strategy of defeat? They cannot be attributed to incompetence. If Marshall were merely stupid, the laws of probability would dictate that part of his decisions would serve his country's interest.

McCarthy, having thus expressed himself about one of the nation's most honored citizens, would have presented a very difficult problem to any Republican candidate for President. But for Eisenhower the problem was doubly difficult. He was irrevocably committed to working for unity and hence to playing down differences of opinion in his own party as much as he could. But as a soldier he had served directly under Marshall throughout the war. In the early stage of the war he had been in charge of the War Plans Division in Marshall's office. In 1942 he had been chosen by Marshall to lead the invasion of North Africa, and thereafter had been given, on Marshall's nomination, the supreme command of Allied forces in Europe. He was more indebted to Marshall than to any other man for his opportunities to achieve greatness. And he had only just returned from his command post in Europe where he had had full responsibility for carrying out the military phase of Marshall's peacetime foreign policy. He had often spoken of Marshall's patriotism and wisdom in glowing terms.

Under prodding from the press, Eisenhower addressed the issue on August 23. "There was nothing of disloyalty in General Marshall's soul," he told the press. "If he was not a perfect example of patriotism and loyal servant of the United States, I never saw one." On this he was unequivocal. But would he support McCarthy, in view of McCarthy's denunciation of Marshall? The transcript reports his remarks as follows:

> I am not going to support anything that smacks to me of un-Americanism—that is un-American in character, and that includes any kind of thing that looks to me like unjust damaging of reputation where the man has not the usual recourse to law. Therefore it is impossible for me to give what you might call blanket support to anyone who holds views that would violate my conception of what is decent, right, just, and fair. At the same time . . . I certainly support those persons who will uproot anything that is subversive or disloyal in the Government. But I think the powers of the Government are ample to do it

without damaging the reputation of any man. Therefore, I will never give blanket endorsement to anyone who has clearly violated what I am talking about: my ideas of American customs and rights.

. . . If a man has been properly nominated by the Republicans in his state, I am going to state clearly that I want to see the Republican organization elected.

The reporters were apparently not satisfied that they fully understood him, and he was asked again specifically whether he would support McCarthy for re-election to the Senate. His reply was, "I will say to you and [*sic*] I will support him as a member of the Republican organization." On September 10, in the Republican primary, McCarthy won a landslide victory for renomination. When asked for his reaction, General Eisenhower replied, "No comment."

On October 3 Eisenhower's campaign train was moving north through Illinois, heading for Wisconsin. Senator McCarthy boarded the train at Peoria. There are no authentic reports of his conversations with General Eisenhower at that time. But there is no doubt that it was at this time that the issue of Eisenhower's text for a speech in Milwaukee was debated. Rumors emanated from the train that Eisenhower was planning to make, in Milwaukee, some extended remarks in praise of General Marshall, and that McCarthy was urging him to delete them. The rumors were officially denied. In any case, the Milwaukee speech, as delivered, contained no reference to Marshall, and Senator McCarthy campaigned with Eisenhower across the state. On October 12, W. H. Lawrence reported in the *New York Times* that Eisenhower had told newsmen on October 5, "off-the-record," that he had deleted the remarks because McCarthy "had suggested there were better places than his [McCarthy's] home state to make that speech," and because Governor Kohler of Wisconsin

had urged him not to use the passage. Eisenhower indicated
that he felt obliged to accept the advice of his "host."

At various stops across Wisconsin, McCarthy appeared on
the platform with Eisenhower and introduced the General at
his home town of Appleton. At Green Bay, Eisenhower made
a direct appeal for McCarthy's re-election, though he attempted to spell out the reasons for his continued hesitation:

> It is, of course, well known, ladies and gentlemen, to you
> and to many others that there have been differences of
> opinion, sometimes on important matters, between me and
> other people in the Republican Party. Indeed, it would be
> a miracle if there were not.
>
> The differences between me and Senator McCarthy are
> well known to others. But what is more important, they are
> well known to him and to me and we have discussed them.
> I want to make one thing very clear. The purposes that he
> and I have of ridding this government of the incompetents,
> the dishonest, and above all, the subversive and the dis-
> loyal are one and the same. Our differences, therefore, have
> nothing to do with the end result we are seeking. The differ-
> ences apply to method.

Agreement on "purposes" but disagreement on "method" was
thus the formula by which the General hoped to resolve his
dilemma. He would support McCarthy for re-election because
McCarthy was the duly nominated Republican candidate,
yet reserve the right to criticize the Senator's methods. But
McCarthy's purposes were never in dispute. No one favored
"incompetents, the dishonest . . . the subversive and the
disloyal" in government or anywhere else. It was McCarthy's
"methods" by which he had made his reputation, and it was
his "methods" which stamped his name on his time. Without
the "methods" there would have been no McCarthy era in

American history and no problem for the Republican candidate to face. Eisenhower's position was well-calculated to keep him free of the issue of McCarthyism by appearing to be on both sides of it. To the liberal, anti-McCarthy wing of his own party he seemed to agree with men like Hoffman that McCarthy's "un-American" methods were intolerable, while to the Old Guard he seemed to support McCarthy as a loyal Republican. If a few of his original supporters, like Senator Morse of Oregon, were disenchanted, to the nation at large he appeared as a champion of civil liberties and fair play at the same time that he appeared to McCarthy supporters to stand shoulder to shoulder with the Senator as a fighter against communism and subversion.

After the Wisconsin tour, Eisenhower did not again speak of McCarthy during the campaign. At Newark, New Jersey, on October 18, he at last spoke his praise of General Marshall. Under continued goadings by Democrats he became angry and nettled:

> I have abandoned no part of my belief in any of the men with whom I have been privileged to work and whom I consider great American patriots. In this group stands General George Marshall.

On the other hand, Eisenhower seems to have made no effort to forestall the speech of Senator McCarthy at Chicago on October 27, nor did he repudiate it afterward. This speech, McCarthy's only appearance on nationwide television during the campaign, was carefully and effectively built up by advance publicity and rumor. The Senator hinted regularly that he had a devastating case against Stevenson's patriotism and that, at the psychological moment, he would reveal it. Perhaps because he did not wish to alter his fixed policy of never mentioning Stevenson's name, Eisenhower did not at any time see fit to include Stevenson among his list of "patriots," or otherwise move to stem the tide of rumor. Or perhaps, under political advice, he decided that his own problem with McCarthy was sufficient without taking on Stevenson's as well. Whatever the reasons, Eisenhower's tactics of disengagement put him in the role of bystander while the poison of slander accumulated in the American atmosphere.

The contrived anticipation of McCarthy's speech was so great that it could not have measured up to expectation. But it was malign enough. "Let me make it clear," he said, "that I'm only covering his [Stevenson's] history in so far as it deals with his aid to the Communist cause and the extent to which he is part and parcel of the Acheson–Hiss–Lattimore group." Armed with "documents" and "exhibits," McCarthy undertook his "case against Stevenson" by associating Stevenson's friends and advisers, like Bernard De Voto, Archibald MacLeish, Arthur Schlesinger, Jr., and others, with liberal organizations which he asserted were either fronts for Communist subversion or whose members, in turn, were members of other organizations which had been designated by various congressional committees as fronts. Stevenson, whom he charged only obliquely with front activities, was branded for his association with his advisers. Much of McCarthy's "material" was drawn from published hearings of the McCarran Senate Subcommittee on Internal Security regarding the Institute for Pacific Relations, and bore no relation at all to the chief object of his attack—Stevenson himself. In its elements McCarthy's "case" contained nothing about either Stevenson or his friends which even remotely indicated communism or disloyalty. But the effect of the whole was to suggest to the unwary that the Democratic candidate was a part of that dreadful conspiracy McCarthy had previously attributed to Marshall and others.

The speech was a classic example of McCarthy's "method"

—the method which Eisenhower had denounced. When it was employed to destroy his opponent, the General was silent. That is, he was silent regarding both McCarthy and Stevenson. But he evidently felt a need once more to assert his adherence to decency and fair play, as Democrats, and some editors, cried "foul" at the tactics of McCarthy and of the General's running mate, Senator Nixon. At Chicago, on November 1, he spoke as follows:

> There has been considerable concern—and rightfully so—about methods to be used in rooting communism out of our government. There are those who believe that any means are justified by the end of rooting out communism. There are those who believe that the preservation of democracy and the preservation of the soul of freedom in this country can and must be accomplished with decency and fairness and due process of law. I belong to this second school.

Thus he continued to commit himself to decency and fairness, but at a level so abstract that what he said could mean whatever his hearers, regardless of their views, wanted him to mean. McCarthy could go on attacking the loyalty and patriotism of Stevenson and others with no fear of reprisal from the leader of the Republican party. But before many months could pass Eisenhower and the nation would pay a heavy price for the President's equivocal attitude.

### III

Throughout the campaign of 1952 Stevenson was, of course, the principal target of McCarthyism. While McCarthy posed no such dilemma for Stevenson as he did for Eisenhower, he nevertheless presented a major problem of political strategy. Stevenson had not only to defend himself personally, but also to defend the record of the Truman administration and of earlier leaders of the Democratic party. Americans were nearly as curious to see how Stevenson would behave in these circumstances as they were to see what Eisenhower would do with McCarthy.

There was a clear indication in his opening remarks to the Democratic Convention, before his nomination. Speaking ostensibly of the Chicago region, Stevenson said:

> Here, on the prairies of Illinois and the Middle West, we can see a long way in all directions. We look to east, to west, to north and south. Our commerce, our ideas, come and go in all directions. Here there are no barriers, no defenses, to ideas and aspirations. We want none; we want no shackles of the mind or the spirit, no rigid patterns of thought, no iron conformity. We want only the faith and conviction that triumph in free and fair contest.

A few days later, in his speech of acceptance, he exhorted his party and the people to be "contemptuous of lies, half truths, circuses and demagoguery." Though these sentiments were expressed with a cutting edge, it was by no means clear whether Stevenson would wait until he was attacked directly by McCarthy and other employers of the McCarthy method or would himself take the offensive. He decided on the latter course. In order to make the greatest possible impact on public opinion he chose to state his position under dramatic circumstances.

On August 27, Stevenson addressed the annual convention of the American Legion in New York City. A large number of McCarthy's followers were members of the Legion, and the Legion itself had been for a long time active in the anti-Communist campaign. To its hundreds of local committees on "un-American activities," McCarthy had become a hero.

So had Democratic Senator Pat McCarran of Nevada, chairman of the Senate Subcommittee on Internal Security. McCarren was less flamboyant than McCarthy and less given to seeking headlines by means of unsupported charges of disloyalty against well-known citizens. But he had been conducting lengthy and well-publicized hearings into alleged subversive activities in various private organizations, and his name was attached both to the restrictive immigration law of 1950 and to the internal security statute of the same year, which had nearly outlawed the Communist party and Communist front organizations. While liberals and moderates were deeply disturbed by the fear and rumormongering encouraged by both Senators, many Legionnaires looked upon them as symbols of patriotism. To the dismay of at least some Democratic professionals, Stevenson undertook the risky, indeed paradoxical, task of reading the Legion a lecture on patriotism:

> We talk a great deal about patriotism. What do we mean by patriotism in the context of our times? I venture to suggest that what we mean is a sense of national responsibility which will enable America to remain master of her power—to walk with it in serenity and wisdom, with self-respect and the respect of all mankind; a patriotism that puts country ahead of self; a patriotism which is not short, frenzied outbursts of emotion, but the tranquil and steady dedication of a lifetime. The dedication of a lifetime—these are words that are easy to utter, but this is a mighty assignment. For it is often easier to fight for principles than to live up to them.

Here was the Stevenson emphasis on responsibility, but the theme was general enough to avoid giving particular offense. His risk, and his claim to leadership, lay in applying the generalization. And Stevenson presently applied it forcefully to both the McCarran and McCarthy brands of "patriotism":

> There are men among us who use "patriotism" as a club for attacking other Americans. What can we say for the self-styled patriot who thinks that a Negro, a Jew, a Catholic, or a Japanese-American is less an American than he? That betrays the deepest article of our faith, the belief in individual liberty and equality which has always been the heart and soul of the American idea.

Stevenson thus included McCarran's nativism while going beyond it in his attack on discrimination, racial and national. Then quickly to McCarthyism:

> What can we say for the man who proclaims himself a patriot—and then for political or personal reasons attacks the patriotism of faithful public servants? I give you, as a shocking example, the attacks which have been made on the loyalty and the motives of our great wartime Chief of Staff, General Marshall. To me this is the type of "patriotism" which is, in Dr. Johnson's phrase, "the last refuge of a scoundrel."

Many in his audience might know little of Dr. Johnson, but there could be no mistaking Stevenson's meaning.

He was not content merely to denounce; he wanted to explain:

> The anatomy of patriotism is complex. But surely intolerance and public irresponsibility cannot be cloaked in the shining armor of rectitude and self-righteousness. Nor can the denial of the right to hold ideas that are different— the freedom of man to think as he pleases. To strike free-

dom of the mind with the fist of patriotism is an old and ugly subtlety.

Thus freedom of the mind should be the essence of patriotism for Americans. Damage to reputation by irresponsible attacks on a man's patriotism is intolerable in itself, but it is worse because it threatens the free mind for everyone.

And the freedom of the mind, my friends, has served America well. The vigor of our political life, our capacity for change, our cultural, scientific and industrial achievements, all derive from free inquiry, from the free mind— from the imagination, resourcefulness and daring of men who are not afraid of new ideas. Most all of us favor free enterprise for business. Let us also favor free enterprise for the mind.

As he turned to the problem of communism, Stevenson the moralist gave way a bit to Stevenson the politician:

Many of the threats to our cherished freedom in these anxious, troubled times arise, it seems to me, from a healthy apprehension about the communist menace within our country. Communism is abhorrent. It is strangulation of the individual; it is death for the soul. Americans who have surrendered to this misbegotten idol have surrendered their right to our trust. And there can be no secure place for them in our public life.

It is reasonable to speculate, at this point, whether Stevenson himself thought that American "apprehension about the communist menace within our country" was "healthy," or whether he thought, rather, that his audience would think it healthy. And his language in denouncing American Communists does not immediately square with his emphasis on the "free mind." But a political leader cannot lead without followers, and Stevenson may ·well have decided on these concessions to popular feeling in order to carry his audience with him to his final, critical point:

> Yet, as I have said before, we must take care not to burn down the barn to kill the rats. All of us, and especially patriotic organizations of enormous influence like the American Legion, must be vigilant in protecting our birthright from its too zealous friends while protecting it from its evil enemies.
>
> The tragedy of our day is the climate of fear in which we live, and fear breeds repression. Too often sinister threats to the Bill of Rights, to freedom of the mind, are concealed under the patriotic cloak of anti-communism.

Thus at the outset of his national career Stevenson left no doubt as to where he stood on the spirit and characteristic methods of McCarthyism. With one calculated stroke he cut himself off from any possible support from McCarthy's followers specifically, and from right-wing anti-Communists and nativists generally. His position was partisan, but it enabled him to assume leadership among those who were already confirmed in their opposition to McCarthyism, in and out of his party. At the same time, by rejecting McCarranism (emphasized a few weeks later by his rejection of McCarran personally as "not my kind of Democrat") he exchanged potential, if not probable, votes for leadership of the growing opposition to that variety of obscurantism.

It may be argued that Stevenson could not have had McCarthyite support in any case. But what is more important, he risked the loss of votes from far larger numbers of people who were either uncommitted or timid. In doing so he gained freedom to lead if he could, to bring whatever influence he could muster to building an eventual effective majority against McCarthy and McCarthyism, McCarran and McCarranism. The high proportion of approval his speech received in the press, including many strongly pro-Eisenhower Republican papers, suggests that such a majority was already stirring. Stevenson, by tactics of engagement, could be as effectively its leader as his capabilities and circumstances permitted, while Eisenhower, by disengagement, was rendering his "crusade for freedom" an ineffectual executive instrument, despite its effectiveness at the polls.

For some weeks Stevenson largely ignored the attacks of McCarthy and others upon himself and upon the Truman administration, while public attention was more centrally focused on Eisenhower's relations with McCarthy. Then on October 8 at Madison, in McCarthy's home state of Wisconsin, he took occasion to quote first Theodore Roosevelt, on the harm that is done to a body politic "by those men who, through reckless and indiscriminate accusation of good men and bad men, honest men and dishonest men alike, finally so hopelessly puzzle the public that they do not believe that any man in public life is entirely straight; while on the other hand, they lose all indignation against the man who is really crooked"; and then Aristotle, that "history shows that almost all tyrants have been demagogues who gained favor with the people by their accusations of the notables." "I would shudder for this country," Stevenson said, "if I thought that we too must surrender to the sinister figure of the Inquisition, of the great accuser. . . . Some, perhaps, find it politically profitable to cultivate the vineyards of anxiety. I would warn them lest they reap the grapes of wrath."

Meanwhile, attempts by Republican speakers to associate Stevenson with persons accused of subversion or of being "soft on communism" became more and more frequent. When the Republican candidate for Vice President, Senator Richard Nixon, added his voice to the chorus, the political situation for Stevenson took on a different character. McCarthy could not be unmistakably identified as a spokesman for Eisenhower, but Nixon would inevitably be so identified.

Richard Nixon was not perhaps in the strict sense a McCarthyite. He had served, as a member of the House, on the Committee on Un-American Activities and had associated himself with the efforts of that committee during the Republican-controlled Eightieth Congress (1947–49) to discover "Communists" in the Roosevelt and Truman administrations. But he was not so much given to making unsupported accusations of prominent persons, and presented himself as a more independent leader in raising a hue and cry about treason. As a candidate for the Senate in California in 1950 he had charged his opponent, Congresswoman Helen Gahagan Douglas, with "softness on communism" and endeavored to associate her with "pink" groups, and he had received strong support from followers of McCarthy.

But Nixon's chief claim to national attention was made in his persistence in exposing the alleged Communist past of Alger Hiss, a former high official of the State Department. He gave strong support to Whittaker Chambers, then an editor of *Time* but formerly a Soviet agent, in Chambers' gradually developed testimony against Hiss. Chambers himself states in his book *Witness* that Nixon did more than anyone else to encourage him to continue his testimony before the House committee at a time when, so Chambers thought, public opinion was overwhelmingly against him and favorable to Hiss. Nixon succeeded remarkably in identifying himself with the exposure of Hiss. Thus as public opinion toward Hiss changed to the view that he was guilty of turning over government secrets to the Russians, through Chambers, and as the jury confirmed that opinion in the second Hiss trial, Nixon appeared more and more in the light of a major hero in the fight against communism. It was common, by 1952, to speak of Nixon as "the man who put Alger Hiss behind the bars."

Now, as candidate for Vice President, Nixon began to capitalize on his popular role in the Hiss case. His approach during the campaign was frankly partisan. While Eisenhower followed a calculated policy of offending as few Americans of either party as possible, Nixon made no effort to win the support of ardent Democrats. Instead, he endeavored to consolidate Republican support for Eisenhower and to undermine whatever confidence independent voters might have in Stevenson. Thus, on October 8, he asserted that four more years of Democratic rule would bring "more Alger Hisses, more atomic spies, more crises." During the middle days of October he began to insinuate that Stevenson might be soft on communism. "Stevenson himself," said Nixon on October 10, "hasn't even backbone training, for he is a graduate of Dean Acheson's spineless school of diplomacy which cost the free world 600,000,000 former allies in the past seven years of Trumanism." Nixon made much of a deposition Stevenson had been called upon to give, in 1949, that Alger Hiss had a good reputation at the time Stevenson knew him. Carefully avoiding any claim that Stevenson was himself a Communist or a subversive, he repeatedly suggested to his audiences that Stevenson's judgment was bad, that he would be a weak President, and that he had demonstrated his weakness by his manner of handling the Hiss matter.

While Nixon employed tactics of insinuation and oblique attack on Stevenson's character, McCarthy was becoming louder and more direct as the time for his Chicago speech approached. Perhaps his most vulgar contribution to the campaign of fear and suspicion was his proposal that if he could board Stevenson's train with a club, he might be able to make a "good American out of him." At this juncture Stevenson felt it necessary to make a direct answer to both Nixon and McCarthy, though it is evident from his answering speech that it was Nixon's role that precipitated the decision to reply.

At Cleveland, on October 23, Stevenson turned on his accusers. His purposes were to defend himself personally and to speak as a partisan leader for those opposing McCarthyism. At the same time he undertook to involve General Eisenhower directly in responsibility for what both Nixon and McCarthy were saying. Thus,

> Now plainly I have no concern about what the junior Senator from Wisconsin has to say about me. As an isolated voice he would be unimportant. But he has become more than the voice of a single individual who thinks the way to teach his brand of Americanism is with a club. This man will appear on nation-wide radio and television as the planned climax of the Republican campaign—as the voice of the wing of the Republican Party that lost the nomination but won the nominee. You will hear from the Senator from Wisconsin, with the permission and the approval of General Eisenhower.

Again,

> Only last week, stung by charges that he had surrendered to the Old Guard, the General said that the decisions in this campaign "have been and will be mine alone." He added: "This crusade which I have taken to the American people represents what I, myself, believe." Crusade indeed!

A moment later he remarked that "the General gave his hand to Senator Jenner of Indiana who had called General George C. Marshall a 'living lie' and a 'front man for traitors' —Marshall, the architect of victory and General Eisenhower's greatest benefactor." Stevenson went on to recall the episode of Eisenhower's eliminating praise of Marshall from his Milwaukee speech. Then,

> If the General would publicly embrace those who slandered George Marshall, there is certainly no reason to expect that he would restrain those who would slander me.

The logic of the argument was nearly irrefutable. Under ordinary circumstances a candidate whose responsibility was thus demonstrated might be expected to pick up the gauntlet. But the voting on November 4 showed that Eisenhower had already succeeded in elevating himself, or had been elevated by popular emotion, to a plane where he could not, regardless of logic, be drawn into this sort of controversy with anyone.

Stevenson next turned to Nixon's attack upon his role in the Hiss affair. After recalling his association with Hiss, the nature of his testimony regarding Hiss's reputation, and the circumstances surrounding the testimony, he moved to turn the moral around and direct it at Nixon:

> At no time did I testify on the issue of the guilt or innocence of Alger Hiss as a perjurer or a traitor. As I have repeatedly said, I have never doubted the verdict of the jury which convicted him.
>
> I testified only as to his reputation at the time I knew him. His reputation was good. If I had said it was bad, I would have been a liar. If I had refused to testify at all, I would have been a coward.

But while the brash and patronizing young man who aspires to the Vice Presidency does not charge me with being a communist, he does say that I exercised bad judgment in stating honestly what I had heard from others about Hiss' reputation. "Thou shalt not bear false witness" is one of the Ten Commandments, in case Senator Nixon has not read them lately. And if *he* would not tell and tell honestly what he knew of a defendant's reputation, he would be a coward and unfit for any office.

The responsibility of lawyers to cooperate with courts is greatest of all because they are officers of the court. And Senator Nixon is a lawyer.

He has criticized my judgment. I hope and pray that his standards of "judgment" never prevail in our courts, or our public life at any level, let alone in exalted positions of respect and responsibility.

Stevenson was at least as blunt in these words as were his opponents. He was drawing a line of political warfare with Nixon which reached its climax two years later. It was a partisan response to a partisan challenge. But precisely because the issue *was* partisan, and clearly drawn, Stevenson's leadership was emphasized rather than diminished, regardless of the price he may have paid in votes.

In the next portion of his Cleveland speech Stevenson tried to expose the fallacy of "guilt by association," as applied to himself, by showing that both Eisenhower and John Foster Dulles, soon to be Secretary of State, had had much closer and much more committed associations with Alger Hiss. But his recital of Dulles' support of Hiss as president of the Carnegie Endowment for International Peace, with Eisenhower's concurrence as a member of the Endowment's board of trustees, was both ineffective politically and a distasteful digression for Stevenson himself:

I would never have believed that a Presidential contest with General Eisenhower would have made this speech necessary.

After the campaign Stevenson more than once spoke with regret of his need to deal at all with the Hiss affair. It seemed to him both irrelevant and degrading. In his concluding paragraphs he tried to salvage something positive from a negative situation:

> But I know and you know that we do not strengthen freedom by diminishing it. We do not weaken communism abroad or at home by false or misleading charges carefully timed by unscrupulous men for election purposes. For I believe with all my heart that those who would beguile the voters by lies or half-truths, or corrupt them by fear and falsehood, are committing spiritual treason against our institutions.

Stevenson's Cleveland speech, and the attacks which precipitated it, are abundant and discouraging evidence of the American political climate in the McCarthy era. The election of 1952 provided no clear-cut decision on the issue. General Eisenhower received monumental support from people on both sides of the issue of McCarthyism, and from the uncommitted. He had so successfully disengaged himself from the controversy that no positive interpretation could be placed upon his victory. The simultaneous congressional elections were a stand-off in which neither party could claim a decisive advantage. McCarthy himself was re-elected but trailed far behind Eisenhower in Wisconsin, and certain leading Republican worriers about communism among Democrats, like Kem of Missouri and Ecton of Montana, were defeated. The controversy, with all its disastrous effects on

American morale at home and American prestige abroad, was to continue for two more years. But Stevenson, in defeat, had nevertheless consolidated a position of leadership among anti-McCarthy forces which enabled him, paradoxically, to play the role a dramatist would certainly have assigned in advance to Eisenhower.

## IV

Their slender margin of victory in the congressional elections of 1952 enabled Republicans to organize the Senate when the Eighty-Second Congress convened in January, 1953. Through seniority, Senator McCarthy became chairman of the Senate's Permanent Subcommittee on Investigations, a post which he soon used to bring the national spotlight into still more direct focus upon himself. While he had been content in previous years to concentrate his attacks almost exclusively on Democrats, he now began to interfere in the conduct of foreign policy, and presently to attack Republicans. By the end of March, President Eisenhower was already face to face with the consequences of his disengagement policy.

On the Senate floor McCarthy and his associates bitterly attacked the President's appointment of Charles E. Bohlen to be Ambassador to the Soviet Union, charging that Bohlen was a member of the same clique of State Department officials whom they had attacked as "soft on communism." The new Republican President was forced to defend his choice of a skilled and veteran career diplomat against Republican attacks and to depend in large measure upon Democrats for support.

On March 29 McCarthy called a press conference to announce that he had "negotiated" an agreement with Greek shipowners to cut their shipping to China between 10 and 45 per cent. Harold Stassen, a 1952 candidate for the Republican presidential nomination who had thrown his delegate support to Eisenhower and was now Director for Mutual Security, protested that McCarthy's action was direct interference in the affairs of his office and violated the separation of powers. After a call at the White House Mr. Stassen muffled his criticism. The President, in his press conference of April 3, said that he did not consider McCarthy's "negotiations" as undermining the foreign policy of the Executive, since McCarthy had no constitutional authority to enter into agreements on behalf of the United States. "He doubted if that would undermine the prestige and power that resided in the Government and its various parts, as fixed by the Constitution."\* While the world press, both friendly and unfriendly, wondered at the President's failure to support his immediate subordinates, McCarthy received no official rebuke from the White House.

Perhaps encouraged by his first venture into foreign affairs, in mid-April McCarthy sent two young employees of his committee, Roy Cohn and G. David Schine, to Europe, to investigate "waste, inefficiency, and loyalty" in American missions. At the same time he held lengthy hearings regarding the personnel of such agencies as the Voice of America, to dramatize his "findings" that Communists of Communist dupes had infiltrated our overseas propaganda program. He charged that American overseas libraries were filled with books by Communist and subversive authors, and urged that these books be withdrawn and destroyed. Apparently stung

---

\* In the earlier period of his administration Eisenhower's statements at press conferences were not quoted directly. All quotations from presidential press conferences are taken directly from the transcripts released by the White House.

to anger by McCarthy's charges and interference, Eisenhower took advantage of an impromptu address at Dartmouth College on June 15 to speak out for freedom of the mind:

> Don't join the bookburners. Don't think you are going to conceal faults by concealing evidence they ever existed. Don't be afraid to go to the library and read every book as long as any document does not offend our own ideas of decency. That should be the only censorship.
>
> How will we defeat communism unless we know what it is . . . what it teaches—why does it have such an appeal for men? Now we have got to fight it with something better. Not try to conceal the thinking of our people. They are part of America and even if they think ideas that are contrary to ours they have a right to them, a right to record them and a right to have them in places where they are accessible to others. It is unquestioned or it is not America.

While the rapidly increasing numbers of McCarthy's opponents at home and abroad were cheering this forthright stand, they were thrown into confusion by the President's press conference only a few days later:

> *Question:* Mr. President, your speech last Sunday at Dartmouth was interpreted or accepted by a great many people as being critical of a school of thought represented by Senator McCarthy; is that right or wrong?
>
> *Answer:* The reporter had been around him long enough to know he never talks personalities, and he thinks that we will get along faster in most of these conferences if we remember that he does not talk personalities and refuses to do so.

Having withdrawn from a direct tangle with McCarthy, Eisenhower went on to qualify his Dartmouth address as follows:

> His speech, he thinks, should stand by itself, but he will amplify to this extent: by no means, when he talks about books or the right of dissemination of knowledge, is he talking about any document or any other kind of thing that attempts to propagandize America to communism.

Thus, the President blunted the issue, and his earlier evasions contributed to McCarthy's unqualified victory. By the end of the month the State Department admitted that it had withdrawn or destroyed over three hundred books on overseas bookshelves by some forty authors—and some of them had been burned.

A detailed recital of McCarthy's activities during the summer and fall of 1953 would be tedious. It is enough to say that his attacks on the Eisenhower administration multiplied at a rapid pace, while the President moved to counter him only once. This was in connection with McCarthy's appointment of J. B. Matthews to the staff of his Senate committee. Matthews had written an article in the July *American Mercury* asserting that the American clergy were heavily infiltrated by Communists and fellow travelers. His appointment by McCarthy aroused strong and nearly unanimous criticism. In response to a telegram from the Commission on Religious Organizations of the National Conference of Christians and Jews, Eisenhower said:

> Generalized and irresponsible attacks that sweepingly condemn the whole of any group of citizens are alien to America. Such attacks betray contempt for the principles of freedom and decency.

In a cause so broadly popular as to include all but a few voices of the extreme right, Eisenhower was an articulate spokesman. Matthews resigned. But it is significant that even on such an issue the President's words were highly generalized, and he did not mention McCarthy.

During the same period, when McCarthy clashed with Under Secretary of State Walter Bedell Smith regarding which books should be allowed in overseas libraries, when he demanded the appearance of top officials of the Central Intelligence Agency before his committee, and when the Department of the Army accused McCarthy of violating national security by releasing a document McCarthy considered "95 per cent Communist propaganda," Eisenhower made no public statements. In October McCarthy began his investigation of Fort Monmouth with claims that he would reveal "ten years of espionage" at that post. It was this investigation which produced McCarthy's monotonous query, "Who promoted Peress?" and his denunciation of General Zwicker, a hero of World War II, as a "disgrace to the uniform." The same hearings led to his final clashes with Secretary of the Army Robert Stevens.

By this time McCarthy's domination of national political news was nearly complete. On November 16 Leonard W. Hall, Republican national chairman, said that communism would be the "big issue" of the 1954 congressional election campaign. On November 25 McCarthy said that communism would be the chief issue of the campaign and that the election would center on his own chairmanship of the Subcommittee on Investigation. On December 3 Eisenhower told his press conference that communism would *not* be an issue in the 1954 campaign, and Secretary Dulles on behalf of the administration rejected the foreign policy proposals McCarthy had made in his November 25 speech. But on December 9 William Knowland, Republican Senate majority leader, stated that communism would be the issue in *both* the 1954 and 1956 elections.

There were increasing signs that the President had entirely lost patience with McCarthy. In December, January, and February he several times gave strong public support to Army Secretary Stevens, who was under constant attack by McCarthy, and made it clear that he joined with both Democratic and Republican leaders in the Senate in wishing to have McCarthy's power to hold "one-man" hearings curbed. But the position of Hall, the support Knowland and other Republican leaders were giving to McCarthyism, and the continuing raucous pronouncements of McCarthy himself indicated that Eisenhower, whatever his personal wishes may have been, was unable or reluctant for political reasons to provide effective leadership either of his party or of the nation as a whole.

Perhaps the most disheartening evidence came with the incident of an implied attack on the patriotism of former President Truman by Attorney General Brownell. When Truman was given an opportunity on nationwide television to refute the suggestion that he had retained Harry Dexter White in high office, knowing that the FBI considered White disloyal, the ex-President not only denounced Brownell's charges but directly attacked the administration for submitting to McCarthyism. Yet it was McCarthy himself who received free time to reply (speech of November 25), and used the occasion to speak not only for himself but for the Republican administration. To the nation and the world it seemed that the past, represented by Truman, and the present, represented by McCarthy, were at issue over treason in the American government. Eisenhower appeared as little more than a bystander. Indeed, he contented himself with a brief statement of confidence in Truman's patriotism, without in any way disavowing Brownell's charges or criticizing

him for making them. But the ranks of McCarthy's opponents were at last forming for decisive action.

## V

During the traditional Lincoln Day celebrations by the Republican party in January and February, 1954, McCarthy
made his final "all out" bid for leadership of his party and
of the anti-Communist independents. In a series of speeches,
arranged by the Republican National Committee, he played
upon the theme of "Twenty Years of Treason," denouncing
the Democratic party as a party of war, crisis, subversion,
and treason. Other Republican speakers followed his example
in cacophonous chorus. While some Republican Lincoln Day
speakers were singing the praises of Eisenhower and talking
little or not at all of communism and treason, their voices
were scarcely audible above the din of McCarthy and his
band, which now included some strange recruits like Governor Thomas E. Dewey of New York, Eisenhower's sponsor
in the Republican party, and well known to be close to the
White House. Dewey, indeed, had sounded the opening note
of the season at Hartford on December 16, where he had
said, for example:

> Remember that the words Truman and Democrat mean
> the loss of 450,000,000 Chinese to the free world. Remem-
> ber that the words Truman and Democrat mean diplomatic
> failure, military failure, death and tragedy.

At another point he had accused Democrats of being "afraid
that the American people will discover what a nice feeling
it is to have a government which is not infested with spies
and traitors."

On March 7, as leader of the Democratic party, Adlai Stevenson decisively intervened. During the greater part of 1953 he had been abroad on a world tour. As an unofficial roving ambassador for his party and as a representative statesman of his country with a wider following and even greater respect abroad than at home, he had felt obliged in his public statements to support President Eisenhower as often as possible. He had observed the disfiguring effects of McCarthyism upon the image of America abroad, and had repeatedly tried to counter by denouncing McCarthy and offering assurances that the Senator represented neither the nation as a whole nor the Eisenhower administration. Thus, to take examples almost at random, in Madras, on May 8, Stevenson answered a reporter's question about McCarthy's "un-American activities":

> I have time and again expressed my views about that. My views have not changed. After all there are queer people in every country. In America we have our queer people, including Mr. McCarthy. America is a free country, and you can give full expression to your views."

In New Delhi, on May 13:

> *Question:* How can Senator Joseph R. McCarthy have such widespread support in a democratic country?
>
> *Answer:* It all depends on what you are talking about when you say he has widespread support. We have found disloyal persons in high places. Senator McCarthy says he wants to root out subversives, and he does have support for that goal. But if you are talking about his means, such as guilt by association and irresponsible accusations, then I do not believe he has widespread support.

Or in London, on July 29:

> There are many people in our country who feel that the
> idea of doing something more to awaken concern for the
> Soviet Communist imperial menace is a desirable thing to
> do. The objection is not so much to the objectives (of
> McCarthyism) as to the methods. I never quite understand
> why they always say he alerted the country to the menace
> of communism. We were alert to it in 1947, as the Marshall
> Plan showed.

Toward the conclusion of his world tour, in an interview
he gave to *U. S. News and World Report* in Paris, Stevenson
answered a question as to the effect McCarthyism had had on
American relations abroad by the one word "bad." In the
following months he frequently expressed the gravest concern
for America's reputation abroad, underlining the point that
at every stop of his tour almost the first question he was asked
had to do with McCarthy.

At home Stevenson had pleaded for an end to character
defamation as a political instrument and urged the President
to use his influence to restrain McCarthy and others of similar persuasion. But events had forced him to conclude unhappily that Eisenhower must be forced either to denounce
McCarthy vigorously and finally, or else take full responsibility for him. Now, at Miami Beach, before the Democratic
National Committee Southern Conference, Stevenson moved
to precipitate the downfall of McCarthy and an eventual end
to the McCarthy era.

His purpose was clear from his opening sentences:

> I do not propose to respond in kind to the calculated
> campaign of deceit to which we have been exposed, nor

to the insensate attacks on all Democrats as traitors, Communists, and murderers of our sons.

Those of us—and they are most of us—who are more American than Democrats or Republicans count some things more important than the winning or losing of elections.

There is a peace still to be won, an economy which needs some attention, some freedoms to be secured, an atom to be controlled—all through the delicate, sensitive and indispensable processes of democracy—processes which demand, at the least, that people's vision be clear, that they be told the truth, and that they respect one another.

Against this perspective of the situation in the United States and the world, Stevenson went on:

It is wicked and it is subversive for public officials to try deliberately to replace reason with passion; to substitute hatred for honest difference. . . .

Recalling the current McCarthy theme of "Twenty Years of Treason," he drew out the irony of associating that theme with Lincoln:

When one party says that the other is the party of traitors who have deliberately conspired to betray America, to fill our government services with Communists and spies, to send our young men to unnecessary death in Korea, they violate not only the limits of partisanship, they offend not only the credulity of the people, but they stain the vision of America and of democracy for us and for the world we seek to lead. That such things are said under the offi-

cial sponsorship of the Republican Party in celebration of
the birthday of Abraham Lincoln adds desecration to
defamation. This is the first time that politicians, Republicans at that, have sought to split the Union—in Lincoln's
honor.

Here Stevenson might perhaps have shifted from irony to
humorous invective. At other times and on other occasions
he did so. But now was a time, unmistakably, for high seriousness:

> This system of ours is wholly dependent upon a mutual
> confidence in the loyalty, the patriotism, the integrity of
> purpose of both parties. Extremism produces extremism,
> lies beget lies. The infection of bitterness and hatred
> spreads all too quickly in these anxious days from one
> area of our life to another. And those who live by the
> sword of slander also may perish by it, for now it is also
> being used against distinguished Republicans. We have
> just seen a sorry example of this in the baseless charges
> hurled against our honored Chief Justice. And the highest
> officials of the Pentagon have been charged with "coddling
> Communists" and "shielding treason." General Zwicker,
> one of our great Army's finest officers, is denounced by
> Senator McCarthy as "stupid, arrogant, witless," as "unfit
> to be an officer," and a "disgrace to the uniform." For
> what? For obeying orders. This to a man who has been
> decorated thirteen times for gallantry and brilliance; a
> hero of the Battle of the Bulge. When demagoguery and
> deceit become a national political movement, we Ameri-
> cans are in trouble; not just Democrats, but all of us.

Next, in a brief paragraph, Stevenson catalogued the
special features of McCarthyism as its poison had spread
through the land:

Our State Department has been abused and demoralized. The American voice abroad has been enfeebled. Our educational system has been attacked; our servants of God impugned; a former President maligned; the executive departments invaded; our foreign policy confused; the President himself patronized; and the integrity, loyalty, and morale of the United States Army assailed.

The point now was to fix the responsibility for these things in unforgettable terms—terms which would require an answer:

> The logic of all this is—not only the intimidation and silencing of all independent institutions and opinion in our society, but the capture of one of our great instruments of political action—the Republican Party. The end result, in short, is a malign and fatal totalitarianism.
>
> And why, you ask, do the demagogues triumph so often? The answer is inescapable: because a group of political plungers has persuaded the President that McCarthyism is the best Republican formula for political success.
>
> Had the Eisenhower administration chosen to act in defense of itself and of the nation which it must govern, it would have had the grateful and dedicated support of all but a tiny and deluded minority of our people.

Why had the President and the administration failed to achieve unity in the party and in the nation? Whatever the reason for inaction, results were devastating. Stevenson cut through to the heart of the matter:

> A political party divided against itself, half McCarthy and half Eisenhower, cannot produce national unity—cannot govern with confidence and purpose. And it demonstrates

that, so long as it attempts to share power *with* its enemies, it will inexorably lose power *to* its enemies.

If Eisenhower, the nonpartisan President, could not plead effectively for national unity under such circumstances, Stevenson, the partisan leader, could:

> Perhaps you will say that I am making not a Democratic but a Republican speech; that I am counselling unity and courage in the Republican party and administration. You bet I am!—for as Democrats we don't believe in political extermination of Republicans, nor do we believe in political fratricide; in the extermination of one another. We believe in the republic we exist to serve, and we believe in the two-party system that serves it—that can only serve it, at home and abroad, by the best and noblest of democracy's processes.

Turning to the question of the "numbers game"—how many "security risks" had been discovered and fired from the government by the Eisenhower administration, Stevenson offered another example of dangerously loose talk for which the President was himself personally responsible. For many weeks, beginning in October, 1953, the White House, the Civil Service Commission, and other agencies had been issuing figures, varying from 1,456 to 2,427, on the number of persons discharged under security proceedings, lumping all kinds of "risks" indiscriminately together. The President's own contribution was the number 2,200. The impression was left uncorrected that most or all of these were loyalty cases. But, said Stevenson,

> the only thing we know for sure is the government's reluctant admission that out of more than two million federal

employees only one alleged active Communist has been found.

. . . The President has said he disapproves of all these goings-on—this slander and deceit, this bitterness and ugliness, these attempts to subordinate a nation's common purposes to a divided party's ambitions. He has said so repeatedly in statements to the press—but the nation's ideals continue to be soiled by the mud of political expediency.

Again, appealing for unity among Democrats, Stevenson, as Democratic leader, could speak for the nation:

The internal crisis makes it all the more urgent that the Democratic party remain strong, responsible, and attentive to the nation's business. I note that no Democrat has charged that the whole Republican party is corrupt merely because three Republican Congressmen in a row have been convicted of defrauding the government. We know that Republicans and Democrats alike want better government —government that measures up to the ideals of a proud people, to the dignity which befits the leader among nations, to the standard we think of as the reward citizens receive from a democracy for which they pay and work and pray and fight, and see their sons die to preserve. Now, more than ever, America must be a citadel of sanity and reason. We live in a troubled, dangerous world where the great issues are peace or war and the stakes are life and death.

After turning for a time to criticism of foreign policy, Stevenson closed with yet another plea for reason and unity:

I hope that we can begin to talk with one another about our affairs more seriously, moderately, and honestly, whether it be our foreign policies or the patriotism of our people and public servants. There has been enough—too much—of slander, dissension and deception. We cannot afford such wastage of our resources of mind and spirit, for there is important work to do which will be done together or not at all. It is for us, all of us, to recapture the great unifying spirit which still surges so strongly through the hearts and minds of America. Let us, as Democrats, resist the ugly provocations of this hour and try to cut the pattern of America's future, not from the scraps of dissension and bitterness but rather from the full, rich fabric of America's ideals and aspirations.

"Let us," in Thomas Jefferson's words, "restore to social intercourse that harmony and affection without which liberty and even life itself are dreary things," and without which, I could add, tomorrow's misfortune will mock today's expectations.

Few speeches in these years have had such instantaneous effect on the opinions and behavior of political leaders. It was too powerful and too forthright to be ignored. Many Democratic leaders were, to say the least, uncomfortable. Some feared that it would backfire on the Democratic party in the coming congressional election, as people resented the direct attack on a beloved President. Others, driven to narrow partisanship by McCarthyism, as Stevenson had foreseen, thought that the McCarthy issue should be left alone to break up the Republican party. Even some Democratic leaders who fully approved the speech asked that their names be withheld from news reports. Such was the fear McCarthy had engendered. But Stevenson's courage was matched by political judgment shrewder than that of his colleagues. The editorial response of the *New York Times* is worth quoting at length:

. . . Mr. Stevenson, as we have learned to expect, spoke as something more than a partisan. He spoke as a conscientious American citizen. When he said that he was "counseling unity and courage in the Republican party" he meant it. No Democrat who also loves his country could wish to see the Republican party ruined by a weak or immoral compromise. Mr. Stevenson charged that a "group of political plungers has persuaded the President that McCarthyism is the best Republican formula for political success." This was not the happiest part of his speech. But much of the rest of the speech implied that there are multitudes of Republicans who believe, as he does, in an essentially non-partisan foreign policy and in a wholly non-partisan policy of decency and respect for individual rights in the conduct of our domestic affairs.

President Eisenhower and Mr. Stevenson, sitting down in private together, would still disagree on elements of policy . . . But on principle the two men would agree. They would stand for a fair hearing for accused persons. They would stand for freedom to think, talk and print.

This speech will have to be answered by some Republican whom the people know and respect. It compels an early and definite decision on the McCarthy issue—which will be awaited with interest.

There was not long to wait. Six days later the *New York Times* summarized in an editorial:

This often seems to be a very patient and credulous nation, but sometimes we get too much of a thing or person. We then arise and in our cheerful and exuberant fashion we abate the thing we don't like and we diminish the person we distrust. Something like this may now be happening to Senator Joseph R. McCarthy of Wisconsin, and to the campaign of misrepresentation, intimidation and self-adulation that he has been carrying on under the pretense of fighting communism.

On Saturday night, March 7, Adlai Stevenson charged that "a group of political plungers had persuaded the President that McCarthyism is the best Republican formula for political success." There was only one good answer to this charge, and that answer was made before the official reply, scheduled for last night, went out over the air. The answer was in action more than in words. On Monday the Republican National Committee, after consultation with the President, designated Vice President Richard M. Nixon rather than Senator McCarthy to reply to Mr. Stevenson. On Tuesday Senator Flanders from the conservative state of Vermont, charged Senator McCarthy with "doing his best to shatter the G.O.P."

On Wednesday, in his press conference, President Eisenhower went further than he had ever gone before in rejecting Mr. McCarthy and most of his works. He commended Senator Flanders for his "services." He indignantly denied Mr. Stevenson's charge that the Republicans are "half McCarthy, half Eisenhower." . . .

In his speech of March 13, Vice President Nixon officially repudiated McCarthy on behalf of the Republican party and the Eisenhower administration:

Men who in the past have done effective work exposing Communists in this country have, by reckless talk and questionable method, made themselves the issue rather than the cause they believe in so deeply.

Thereafter McCarthy's downfall was swift and sure. He held the spotlight from April 22 to June 18, during the notorious Army–McCarthy hearings, but he behaved more and more, before his nationwide television audience, like a cornered man. His challenges to the administration were at last emphatically rebuffed, and his "case" collapsed in ever wilder charges and hopeless confusion of detail. On June 12 Senator Flanders moved in the Senate to strip McCarthy of his powers. On July 20 Senator Flanders called upon his Republican colleagues to join in a formal resolution censuring McCarthy's conduct. Yet when Senator Knowland, Republican Senate leader, opposed the Flanders resolution, President Eisenhower once more pathetically wavered:

> *Question:* Does Senator Knowland speak for the Administration in his opposition to the Flanders resolution?
> *Answer:* They hadn't even asked him about it, they hadn't even asked him a thing about it. He had taken no stand whatsoever.
> *Question:* You are taking none now?
> *Answer:* None now.

On August 3 the Senate voted to set up a six-man panel to report on the matter of censuring McCarthy. In his press conference of August 5 the President refused to comment on the Senate proceedings. "That was their business," he said. On September 28, as the congressional election campaign was beginning, the Senate panel unanimously recommended the censure of McCarthy. Full Senate debate was postponed until after the election.

At Indianapolis on September 18, Stevenson, as leader of the Democratic party, made his opening speech of the campaign. Against the background of ideas and events surveyed in this chapter it will be enough to cite the main portion of his brief speech without comment:

The fact is that we are Americans, first, last, and always, and may the day never come when the things that divide us seem more important than the things that unite us. We have many differences with the Republicans on specific issues of national policy, and we want to discuss and debate them because we think they are important. But I hope we may never forget that we hold far more in common with our friends, the Republicans, especially Republicans like Wendell Willkie, than we hold in dispute. Were it not so, neither party could govern, for government rests less on majorities at the polls than on the abiding unity, good sense, and obedience of the people.

Even in these sobering times it would be too much to hope, I suppose, that there might be an end to extravagant claims that one party represents all that is good and the other all that is evil. And we know that shrill voices filled with bitterness and hate have already been raised in our land. A strange, and, it seems to me, truly un-American violence has stained too many utterances in recent months. It was the Republican Governor of New York, twice his party's candidate for President, who damned all Democrats for all time with words too ugly to repeat and too grotesque to believe. It was the Republican Attorney General of the United States who impugned the very loyalty of a former President of the United States—a man who has done more to combat Communism at home and abroad than all the Republican politicians put together— Harry S. Truman. It was a Republican Senator from this great state of Indiana who described Democrats as betrayers. And it was the Republican National Committee itself which sponsored, in memory of Abraham Lincoln, the slogan "Twenty Years of Treason" to describe the two great Democratic decades.

Now this, of course, is not the language of reasoned

political debate. This is the language of clan warfare, of civil war, of flaming passions and unreason. And it is more dangerous than just the debasement of our political dialogue and our political morals, because as it exploits it also aggravates the unhealthy national mood of fear and suspicion of one another that has so hampered the unemotional discussion on which wise public policy must be based; a mood that has so dangerously diverted us from the main jobs of establishing sane foreign policies and evolving sound domestic programs.

We shall hear more and more of these unscrupulous, shrill voices before the people must judge in November. What will the response be? Will the people reject or applaud those who do not even hesitate to recklessly divide America into ugly, bitter factions? I think the good sense of the American people will prevail and that America has already made its decision on these demagogues who rely on defamation, deceit, and double-talk. At any rate, whatever the provocation, we must not be guilty of contributing to irreconcilable divisions in our country and to political delinquency. And, whatever the provocation, I hope and pray that we Democrats will both recognize and respect the difference between cynical politics and principles, between ruthless partisanship and patriotism.

Now I have spoken seriously, indeed piously, about this because the preservation and strengthening of America requires above all the preservation and strengthening of our mutual trust and confidence. No election victory is worth the damage of these central elements of our strength. Weakness begins at home, in doubts, suspicions, and whispers, and if the spirit of America is enfeebled, it will be the result of self-inflicted wounds.

So I say let us dispute our honest differences honestly and let the people decide them on the merits, but let us

Democrats at least not be guilty of sowing discord, mistrust, and hate in this lovely land. As bearers of an honorable and ancient political tradition let us so conduct this momentous campaign as not to weaken but to strengthen the nation in this troubled hour.

Such was the keynote of the Democratic campaign.

Vice President Nixon led the campaign for the Republicans. With McCarthyism doomed and Eisenhower publicly opposed to making "communism" the campaign issue, Nixon was content to rely on insinuation. Here are some examples:

"Trumanism" had been rejected two years ago because it showed hopeless inability to deal with the fourheaded monster that was Korea, Communism, corruption and controls. (September 16)

The Eisenhower Administration happens to believe that when American boys fought and died fighting communism overseas we ought to deal with it effectively here. (September 17)

The Democrats failed to take the advice and heed the warning of the man who knows more about it [the domestic Communist menace] than anybody else, J. Edgar Hoover. This Administration is taking the advice and cooperating with J. Edgar Hoover. (October 5)

The Communist Party is right when it says the 1954 elections are crucial in determining the path America will take. It has determined to conduct its program within the Democratic Party. . . . The previous Administration's lack of understanding of the Communist danger and its failure to deal with it firmly abroad and effectively at home has led to our major difficulties today. The previous Administration unfortunately adopted policies which were soft, vacillating, and inconsistent in dealing with the Communists. (October 22)

Mr. Stevenson has not only testified for Alger Hiss, but he has never made a forthright statement deploring the damage that Hiss and others like him did to America because of the politics and comfort they received from the Truman Administration and its predecessors. (October 24)

If the Democrats win control of Congress it will give a tremendous boost to the Left-Wing elements. (October 25)

Mr. Stevenson has been guilty, probably without being aware that he was doing so, of spreading pro-Communist propaganda as he has attacked with violent fury the economic system of the United States and has praised the Soviet economy. (October 28)

President Eisenhower's statement, in his nationwide address from Denver, that a Democratic victory in the election would mean a "cold war" between Congress and the President, may be dismissed as campaign politics. But his withdrawn attitude toward the question of whether Nixon and others had his approval in their insinuation of Democratic "softness on communism" revealed at the least his detachment from the issues before the country. Thus in his press conference of October 28:

*Question:* Now of late, the Republican leaders who have been campaigning around the country, with the exception of yourself, Sir, have seemed to shift emphasis from the accomplishments of the Congress and your legislative program to the Communist issue. I would like to know, Sir, first, whether they have consulted you on that decision and second, whether it has your approval?

*Answer:* He (the reporter) based his whole question on a statement of what he (the reporter) said appeared to be a Republican attitude at the moment.

He had not read the speeches. He had listened lately to two or three talks here in town, and he had not heard the word "Communist" mentioned. He meant they had originated here in town.

As far as he was concerned none of these people had come to him about the details of their talks. They knew what he believed, and they were going out doing their best in their own way and, he supposed, answering questions or attempting to answer or to present the case as they saw it. But he couldn't possibly comment in detail on the whole generality.

During the final days of the campaign Eisenhower was persuaded to enter actively into the Republican effort. He spoke specifically in behalf of senatorial candidates Ferguson, Cooper, Meek, Bjornson, Bender and Warburton, among others generally endorsed, and made speeches at Detroit, Louisville, Cleveland, and Wilmington, Delaware. Ferguson, Meek, Bjornson, and Warburton were defeated on election day.

In 1952, and again in 1956, Adlai Stevenson, partisan leader and candidate, could not match the popularity of a nonpartisan hero and President. When the choice was for the Presidency, the people seemed to be trying to reach above the issues which divided them to choose as their national symbol a man detached and disengaged. But at the level where the issues were faced and debated, Stevenson, precisely because he was committed, could articulate the thoughts of the growing majority and show the way toward unity. In 1954 he led his party back to power in Congress, and through the election brought an end to the McCarthy era in American life. On December 3, one month after the election, the Senate voted 67 to 22 to condemn the conduct of Senator McCarthy. The triumph belonged to the people, yet their authentic spokesman was Adlai Stevenson, not Dwight Eisenhower.

CHAPTER III

# Civil Rights: Law,
# Conscience, and Leadership

I

ON MAY 17, 1954, while the nation was engrossed in the sickly spectacle of the Army–McCarthy hearings, the Supreme Court handed down its decision in the case of *Brown* vs. *Board of Education* (Topeka, Kansas). Chief Justice Warren, speaking for a unanimous Court, concluded:

> . . . that in the field of public education the doctrine of "separate but equal" has no place. Separate educational facilities are inherently unequal. Therefore, we hold that the plaintiffs and others similarly situated for whom the actions have been brought are, by reason of the segregation complained of, deprived of the equal protection of the laws guaranteed by the Fourteenth Amendment.

Within two years Americans were to be at least as confused and agitated by the meaning of this decision and the events which followed from it as they had been by the storm over McCarthyism. The decision that segregated public schools are unconstitutional, together with the decrees of 1955 ordering gradual desegregation, marked a revolutionary turn in the long struggle of American Negroes to achieve equality with whites. The issue of civil rights, central in American life since the Civil War, took on a new and pressing significance. McCarthyism gave way to a new challenge for all the people, and especially for the President and the leader of his opposition.

For Dwight Eisenhower, leader of the Republican party, civil rights was an inviting and honorable issue. The Republican party was born in the struggle to free the slaves and in its platforms had always championed the rights of Negroes. While Republicans had tended for many years to play down civil rights in practice in exchange for an effective coalition with Southern Democrats in Congress, the new significance given to the issue by the Supreme Court presented Eisenhower with an unparalleled opportunity to revive Republican militance. But his remarkable achievement of carrying on the shoulders of his popularity the electoral votes of four southern states in 1952 seemed to underscore the risks he would run in such a revival. Thus, for Dwight Eisenhower, as nonpartisan President, civil rights could only pose an almost unmanageable political problem.

For Adlai Stevenson civil rights afforded an equally fearful dilemma. His task was to lead a national party which contained largely within itself the issue of equal treatment for Negroes. The Democrats had been, since 1877, in nearly complete control of the southern states, where segregation was the rule, while their chief electoral strength in the twenty years before 1952 had been in the great industrial cities of the North, where Negroes were advancing toward first-class citizenship.

Now, after the passing of McCarthyism, the new questions were—how would Eisenhower, the symbol of peace and non-involvement, deal with the forced issue of desegregation?—and how would Stevenson, the partisan politician, seek to reconcile the radically opposed wings of his party and offer leadership to the nation?

In the campaign of 1952 General Eisenhower successfully avoided controversy on specific questions of civil rights. He was content to speak in broad terms of freedom and equality for "all Americans." His reward was a showing of political strength in the South unprecedented in Republican history. However, early in his first term he began to make cautious claims to advances in civil rights. Thus in a speech to the Young Republicans on June 11, 1953:

> We have taken substantial steps toward ensuring equal civil rights to all our citizens regardless of race or creed or color. . . .
>
> Action has been taken in Army camps and schools. And in the District of Columbia, before the bar of the Supreme Court, the Attorney General has successfully appealed for the upholding of laws barring segregation in all public places in our National Capital.

Another important move was the appointment of Vice President Richard Nixon to chair a committee to oversee the government's policy against discrimination in plants and factories doing contract work for the government.

But none of these steps involved a new departure. President Truman had issued the historic executive order to desegregate the armed forces, and much progress had been made before he left office. The policy against discrimination on government contracts had been inaugurated by President Roosevelt many years before. And some of the cases regarding desegregation in the District of Columbia had been commenced in the previous administration. Eisenhower showed no clear desire to make civil rights an issue. His apparent affinity for southern conservatives, his dependence on the support of northern liberal Democrats in both houses of Congress for the enactment of much of his legislative program, and his favorite posture of nonpartisan detachment combined to make a forthright stand on civil rights nearly impossible.

It was therefore not surprising that his initial reaction to the school desegregation decision was noncommittal. He conferred with the commissioners of the District of Columbia and urged them to make Washington a "model" of the forthcoming transition. But he hastened to agree (May 19) with Governor James F. Byrnes of South Carolina, who had said, "Let's be calm and let's look this thing in the face." There was no comment on Governor Byrnes's threat that his state would abolish public schools before it would desegregate them. In a message to the National Association for the Advancement of Colored People on June 29, Eisenhower spoke of the court's decision as a "milestone of social advancement," but gave no clear indication that he favored it, nor any indication at all that he was concerned about enforcing it. On August 12, at his press conference, he was asked whether he had considered asking Congress for legislation to help him enforce the desegregation program. He answered that the "subject had not even been mentioned to him."

On other matters in the field of civil rights Eisenhower was almost completely silent. The Roosevelt and Truman plans for outlawing racial discrimination in employment practices through federal statute were ignored. Though there was some talk of legislation to enforce voting rights and some tentative proposals were made for a commission to investigate the condition of civil rights generally, it was not until 1957, after his decisive re-election, that Eisenhower proposed a full-scale civil rights bill to the Congress. A "strong" civil rights plank, drafted at the Republican National Convention in September, 1956, was revised, after consultation with the White House, to avoid emphasizing federal enforcement of the desegregation orders of the Supreme Court.

## II

Adlai Stevenson was drawn into the center of national Democratic party affairs in 1952, at a time when civil rights threatened to split that party irreparably. In 1948 northern liberals, with the support of President Truman, had insisted on so strong a civil rights program—especially fair employment practices legislation—that many southern delegates walked out of the nominating convention, and a third party was formed in the South by dissident Democrats. Four states gave their electoral votes to Strom Thurmond, "Dixiecrat" candidate from South Carolina.

While Truman's program failed of enactment in the hands of the traditional coalition of Southern Democrats and conservative Republicans, he and his supporters were insistent that the Democratic party should present a bold face on the issue in 1952. Averell Harriman of New York and Richard Russell of Georgia were candidates for the presidential nomination who dramatized the split in the party. The issue was joined at the convention on the question of requiring an affirmation of "loyalty" whereby Southern Democrats would guarantee to support the nominees of the convention and ensure that their names would appear on the party ballot in their states—an affirmation deemed necessary by northerners, since Dixiecrats had appropriated the party's place on the ballot in several states in the 1948 election. After heated debate there was a dramatic roll call on the seating of certain recalcitrant southern delegations. When the big Illinois delegation voted to seat all duly elected delegates, a gentleman's agreement was reached, avoiding the "oath" but assuring that the nominees would appear on the Democratic ballot, and the delegates were seated. Neither the liberal nor the Dixiecrat wing of the party was satisfied, but the convention was nevertheless able to proceed. There was immediate speculation that the Illinois delegation had acted under instructions from Stevenson. The leader of the delegation, Colonel Jacob Arvey, has repeatedly denied that there was any communication between the delegation and Stevenson. It was nevertheless clear to everyone that their action reflected Stevenson's wishes. From the moment of his nomination a few days later, his policy as leader of the party was consistently in favor of party unity on the best terms obtainable for civil rights.

Thus in an important sense, Stevenson's nomination in 1952 was a compromise negotiated by northern and southern leaders of the Democratic party. That a "band wagon" movement to draft him had developed among rank-and-file delegates, to a point at which, perhaps, it could not have been stopped in any case, should not blur the significance of his nomination in regard to civil rights. Stevenson was not committed to the civil rights position of any group in the party. The convention appeared to think of him as a "moderate." He was, in short, free to struggle in his own way with the dilemma any Democratic presidential candidate must face.

But if Stevenson's previous views on civil rights were not well known, they could easily have been discovered. At the Democratic Convention of 1948, as a member of the credentials committee, he had vigorously opposed the seating of the Mississippi delegation, whose members had announced in advance that they would not support President Truman or any other candidate with Truman's views on civil rights. Stevenson's unsuccessful attempt at that time to gain recognition from Senator Alben Barkley, who was chairing the convention, attracted national attention. Four years later there were a good many delegates from both North and South who remembered. As Assistant to the Secretary of the Navy during World War II, Stevenson had made the first moves toward breaking down color lines in the Navy by encouraging the commissioning of Negro officers. After 1948, as governor of Illinois, he had forbidden discrimination on contracts with the state, had vigorously supported the desegregation of public schools in southern Illinois, had demanded a fair employment practices law, and had not hesitated to use force to end race riots in Cicero. But when he had acted for civil rights at the 1948 convention, he was candidate for governor of a Northern state with many thousands of enfranchised Negroes; when he moved to commission Negro officers in the Navy there were no political consequences for him personally to face; and when he acted for civil rights in Illinois he was on the popular side of the issue. Now that he was the national spokesman of the Democratic party, badly needing the electoral votes of the South, the circumstances were different. If Stevenson were drafted at least partly because he was thought to be a suitable compromise candidate, would he prove to be a "compromiser"?

Stevenson's treatment of the civil rights question turned out almost immediately to be closely parallel to his handling of McCarthyism. In one of his first 1952 campaign speeches, at the New York Democratic Convention on August 28, he tackled the subject with vigor. After inviting the Republicans to debate certain major issues with him "on the plane of serious, factual discussion," he said that civil rights was one such issue:

> The phrase civil rights means a number of concrete things. It means the right to be treated equally before the law. It means the right to equal opportunity for education, employment and decent living conditions. It means that

none of these rights shall be denied because of race or color
or creed. The history of freedom in our country has been
the history of knocking down the barriers to equal rights.
One after another they have fallen, and great names in our
history record their collapse: the Virginia Statute of Religious Freedom, the Bill of Rights, the Emancipation
Proclamation, the Woman's Suffrage Amendment, down to
the 1947 Report of the President's Commission on Civil
Rights.

The record of our progress is a proud one, but it is far
from over. Brave and important tasks remain. We cannot
rest until we honor in fact as well as word the plain language of the Declaration of Independence.

That he was talking in New York City, where the Harlem
Negro vote might be crucial in the election, seemed to be
uppermost in his mind when he shortly spoke such characteristic words as these:

In this discussion of all discussions, let us not be self-
righteous. Let us work for results, not just empty political
advantage. We are dealing here with fundamental human
rights, not just votes.

The moderate tone with which Stevenson proceeded to
develop the theme that civil rights is a national—not merely
a sectional—problem was received without enthusiasm by
northern liberal members of his party.

This is a job for the East, the North and the West, as well
as for the South. I know. I have been a Governor of a
great Northern state. I have had to stop outrages committed
against peaceful and law-abiding minorities. I have twice
proposed to my legislature a law setting up in our state

an enforceable fair employment practices commission. I am very proud to say that the Democrats in our legislature voted almost solidly for the bill. But I must report in simple truth that the bill was lost in Springfield, Illinois, because of virtually solid opposition from the party which claims descent from Abraham Lincoln. All the same, gratifying progress has been made in Illinois toward elimination of job discrimination by the initiative of business itself. And I would be less than fair if I didn't acknowledge it gratefully.

This was moderation, but it was partisan moderation—not avoidance of the issue. His purpose was to lay down a line on civil rights that he could consistently maintain in any section of the country. Thus when he spoke in New York of the South, he admitted the seriousness of the problem there but emphasized southern progress:

In saying this is not a sectional problem, I do not mean to say that there is no particular problem in the South. Of course there is a problem in the South. In many respects, the problem is more serious there than elsewhere. But, just as it is chastening to realize our own failures and shortcomings in the North, so is it both just and hopeful to recognize and admit the great progress in the South. Things are taking place in the South today that would have seemed impossible only a few years ago. In the last two years alone ten state universities have admitted Negro students for the first time to their graduate and professional schools. And that is only one of many examples that could be cited of the wonders that are working in the South.

If such emphasis on gradual southern progress might cost votes in northern cities, it nevertheless suggested the lines upon which future advances might be anticipated, and articulated the hopes of all but extremists in all parts of the nation.

Turning to the specific issue of fair employment practices—an issue on which his party had nearly foundered more than once—he steered a careful course of balanced emphasis on state and federal responsibilities. Thus:

> In the case of equal opportunity for employment, I believe that it is not alone the duty but the enlightened interest of each state to develop its own positive employment practices program—a program adapted to local conditions, emphasizing education and conciliation, and providing for judicial enforcement. That is the kind of law I proposed in Illinois.

If this was a concession to "states' rights," in and out of the Democratic party, the approach was none the less positive. Because it was positive it gave stronger meaning to his proposals for federal action:

> Personally, I have been much impressed by a bill recently reported favorably by the Senate Labor Committee. Only three members opposed it, one of whom was Senator Richard Nixon. Both your New York Senators joined in sponsoring the bill [Lehman, Democrat, and Ives, Republican].
>
> It creates a Federal Commission and encourages it to stay out of any state with an effective commission; by the same token, however, it encourages the states to act because, if they do not, the national government has the power to do so. The bill requires the Federal Commission to undertake a non-partisan and nationwide educational program to proceed by persuasion as far as possible, and,

in cases of complaints of violation, to proceed by very careful deliberation and full and fair hearings. Enforcement
would be by order of a court, not an administrative body.

If there was compromise here it was compromise on methods
of improving the condition of civil rights, not a compromise
with civil rights themselves. Stevenson approached the whole
problem in a spirit of conciliation and moderation, but he
did not hesitate to commit himself.

In less than a month Stevenson was speaking in Richmond,
Virginia. Just as he had chosen the convention of the American Legion to state his position on the pseudo patriotism of
McCarthy and McCarran, so he chose the old capital of the
Confederacy to tell the South he could not tolerate unequal
treatment of minorities. And, again, as he had tried to carry
the Legionnaires with him by conciliation and explanation, so
he now tried to hold, not alienate, his southern fellow citizens.
Reaching below the level of emotion and prejudice, he identified a chief cause of racial discrimination in the elemental
fact of poverty:

> One thing that I have learned is that minority tensions
> are always strongest under conditions of hardship. During
> the long years of Republican neglect and exploitation, many
> Southerners—white and Negro—have suffered even hun-
> ger, the most degrading of man's adversities. All the South,
> in one degree or another, was afflicted with a pathetic lack
> of medical services, poor housing, poor schooling, and a
> hundred other ills flowing from the same source of poverty.
> The once low economic status of the South was produc-
> tive of another—and even more melancholy—phenome-
> non. Many of the lamentable differences between Southern
> whites and Negroes, ascribed by insensitive observers to
> race prejudice, have arisen for other reasons. Here eco-

nomically depressed whites and economically depressed Negroes often had to fight over already-gnawed bones. Then there ensued that most pathetic of struggles: the struggle of the poor against the poor. It is a struggle that can easily become embittered, for hunger has no heart. But, happily, as the economic status of the South has risen, as the farms flourish and in the towns there are jobs for all at good wages, racial tensions have diminished.

It might be good politics to blame racial discrimination on Republicans, but northern Democratic politicians were dismayed that Stevenson was talking of racial tensions at all in such a place and at such a time. They were shocked by what came next:

In the broad field of minority rights, the Democratic Party has stated its position in its platform, a position to which I adhere. I should justly earn your contempt if I talked one way in the South and another way elsewhere. Certainly no intellectually dishonest Presidential candidate could, by an alchemy of election, be converted into an honest President. I shall not go anywhere with beguiling serpent words.

No one could miss the point. The Democratic platform had called for equal treatment regardless of race and for federal action in the field of employment practices. Southern Democrats had fought both the "strong" draft of the plank and the more moderate final version. Many had stated quite simply that they could not support such a plank at all. Stevenson, at Richmond, endorsed it without qualification. Whether this was the reason, or some other, on election day he became the second Democrat since the Civil War to lose Virginia— the other was Al Smith.

But this was not all. He was intent on leaving no doubt
in the minds of southerners that he could be sympathetic
and understanding, conciliatory and moderate, but that he
would not condone racial injustice:

> So long as man remains a little lower than the angels, I
> suppose that human character will never free itself entirely
> from the blemish of prejudice, religious or racial. These are
> prejudices, unhappily, that tend to rise wherever the mi-
> nority in question is large, running here against one group
> and there against another. Some forget this, and, in talking
> of the South, forget that in the South the minority is high.
> Some forget, too, or don't know about the strides the South
> has made in the past decade toward equal treatment.
>
> But I do not attempt to justify the unjustifiable, whether
> it is anti-Negroism in one place, anti-Semitism in an-
> other—or for that matter, anti-Southernism in many places.
> Let none of us be smug on this score, for nowhere in the
> nation have we come to that state of harmonious amity
> between racial and religious groups to which we aspire.

To take such a position on civil rights was, for a Democratic candidate, a calculated risk that was almost certain to
fail at the polls. That it did in large measure fail was demonstrated by the campaign itself and by the counting of the
votes in November. Some southern congressmen and other
Democratic leaders openly repudiated Stevenson and gave
their support to Eisenhower. Still others, including some of
the most prominent, "sat on their hands." Eisenhower carried Virginia, Florida, Tennessee, and Texas, as well as most
of the border states. In the North the Democratic majority
declined in almost all cities with large Negro populations.
While liberal Democratic candidates for Congress were
elected in northern cities, Stevenson generally ran well be-

hind them. It may be argued that Eisenhower's personal popularity was the determining factor, not Stevenson's views. But there can be no doubt that Stevenson's moderate stand was denounced by many southerners as too "radical" and by many northerners as too much of a "compromise." Yet it is equally certain that his stand on civil rights counted immeasurably in the ultimate success of his 1956 campaign for renomination. Above all, he articulated a means of conciliation within the Democratic party which led directly to the enactment in 1957, by a Democratic Congress, of the first civil rights measure since the Reconstruction period, and left Stevenson himself morally and politically prepared to speak for the nation on the crisis of integration in the public schools.

## III

As the Presidential campaign of 1956 approached, Eisenhower remained detached from the civil rights problem. Thus in February, 1955, a bill dealing with the military reserve forces and the National Guard was killed by the coalition of Southern Democrats and conservative Republicans in congressional committees because it had been amended to outlaw segregation; yet the President told his press conference that the record of his administration on desegregation was very good relative "to any other administration I know of," though he was "opposed" to a desegregation rider on the reserve forces bill. When the Supreme Court issued its decree regarding the procedure for desegregating the public schools, May 31, 1955, the President had nothing to say.

When the Lincoln Day period of 1956 came around, there were signs that some Republicans wished to make an election issue of civil rights, particularly the school integration question. Vice President Nixon, for example, in his New York

speech of February 12, praised the decisions of the Supreme
Court on civil rights and spoke of Chief Justice Warren as a
"Republican" Chief Justice. In his press conference the next
day President Eisenhower carefully disowned Nixon's phrase,
saying that when a man goes on the Supreme Court he no
longer has "a political designation." But he did not commit
himself on the theme of Nixon's speech.

On February 29, 1956, President Eisenhower announced
his decision to seek a second term. With the Republican candidate thus determined, national attention was quickly focused
on the Democratic nomination race. Stevenson, who had announced his candidacy some weeks earlier, now became the
chief target of questioning and pressures on civil rights and
the school integration problem. While Eisenhower continued
to avoid the issue, Stevenson could not if he had wished to,
since he was involved in a tense campaign for votes in primary
elections in such states as Minnesota, Oregon, Florida, and
California.

Civil rights presented Stevenson with political problems of
great and pressing complexity and with a heavy burden of
responsibility. As a candidate for the Democratic nomination
he needed the support of a majority of the convention delegates. These could be obtained by ignoring the southern states
and concentrating on eastern, northern, and western states.
Once having won the nomination, he would need all the
Democratic votes he could get from any region, as well as
very large numbers of independent votes. Thus the easy
road to nomination might cost him his best chance to win the
election. On the other hand, he was counseled by many
northern Democrats to rely on the North, not only for the
nomination but for the election, on the ground that a highly
popular stand on civil rights, among other issues, might produce a vote that would more than offset losses in the South.
But this line could well split the Democratic party and would

certainly influence the integration issue. And Stevenson was not only a candidate, he was leader of the Democratic party. To be free to win the nomination by taking the popular civil rights stand in the North, and to plot his strategy for the election in the same way, he would have to abdicate his party leadership. Yet, since the Democratic party, as a national party, contained within itself the segregation issue which the next President would have to face, to give up his leadership in order to smooth the way to the Presidency would, Stevenson was convinced, disqualify him for the office. As a candidate he must not only plead for votes but seek to make his campaign itself a contribution toward resolving the issue of integration. Conscience thus combined with circumstance to bring Stevenson to the crucial decision that he would retain his party leadership and campaign on a firm but "moderate" civil rights platform. Because the President remained uncommitted, Stevenson's decision in effect meant that he would try, as a candidate, not only to lead his party, but to lead the nation. His success within his party won him overwhelming victory at the convention, pulled his party toward unity and a more liberal view of civil rights, and pointed toward enactment of the Civil Rights Law of 1957. While Eisenhower drifted toward the disaster at Little Rock, Stevenson articulated the conscience of the nation.

Stevenson's campaign on the civil rights issue began somewhat uncertainly. In Los Angeles, February 7, he drew groans of disapproval when he told an audience which included many Negroes that he would not enforce desegregation of public schools by the use of federal troops. At the same time he stated his opposition to the Powell Amendment, which would have forbidden federal aid to segregated schools, whether or not such schools were in defiance of court orders to desegregate. Poor schools, he continued, were a major cause of the problem of racial discrimination and he could not favor keeping them poor. Someone in the audience was heard to say, "He's a phoney." He denied that he was "appeasing the South" to win votes for the nomination, insisting, rather, that North and South must live together—that a "Balkanized America" was an unthinkable idea. He pleaded for understanding and patience, arguing that education would provide the only sure means of reaching peaceful solutions to civil rights questions. Pressed to set a target date for the completion of school integration, he suggested January 1, 1963, the centennial anniversary of the Emancipation Proclamation. The audience groaned again. Someone in New York sent him a campaign contribution "to be cashed in 1963."

Such going was rough and pressures on Stevenson mounted, but he could not and would not change his views. What he could and did do was to clarify and expand them. At Hartford, Connecticut, on February 25, in a carefully prepared address, he spoke at length of civil rights.

We sometimes forget, when phrases become familiar, all that they mean. And their real meaning must be in terms of the problems each generation faces.

They mean in terms of the most crucial issue within this beloved land of ours today: equal rights for all— regardless of race and color.

America is nothing unless it stands for equal treatment for all its citizens under the law. And freedom is unfinished business until all citizens may vote and live and go to school and work without encountering in their daily lives barriers which we reject in our law, our conscience and our religion.

Moving from general proposition to particular problem he proceeded to the integration decisions:

The Supreme Court has reaffirmed this essential doctrine of democracy in the school desegregation cases. These decisions speak clearly the law and the conscience of this land. But they recognize that a time for transition and compliance is necessary, time for the adjustments that have to be made. But they do not recognize or permit repudiation of these decisions of the court and of the people.

By asserting that the court spoke the conscience of the land, and that the decisions were those not only of the court but "of the people," Stevenson disentangled the issue from legal snarls and loopholes. The integration decisions were to be accepted not only because the Supreme Court was supreme in the legal system, but because the decisions had popular sanction. He would be moderate, as the court was moderate, but he would not appease or equivocate.

Turning to "interposition," he recalled that Andrew Jackson, a Democratic President, was the first who had to deal with this "legal" device of the states to subvert the national law. He quoted Jackson:

"I consider then," Jackson wrote, "the power to annul a law of the United States, assumed by one State, incompatible with the existence of the Union, contradicted expressly by the letter of the Constitution, unauthorized by its spirit, inconsistent with every principle on which it was founded, and destructive of the great object for which it was formed."

That was essential Democratic doctrine—and American doctrine—120 years ago. It is essential Democratic—and American—doctrine today.

While Stevenson, a candidate for the Democratic presidential nomination, thus repudiated the express views of

many powerful Democrats and the formal resolutions of
several Democratic state legislatures, President Eisenhower
found it impossible to respond unequivocally to the conflicting pressures of people divided on the issue but united
in affection for himself:

> *Question:* As you may know, four of the Southern state
> legislatures have passed interposition resolutions stating
> that the Supreme Court decision outlawing segregation
> has no force and no effect in their state; and I was won-
> dering what you thought about this concept of interposi-
> tion, and what you thought was the role of the Federal
> Government in enforcing the Supreme Court decision?
>
> *Answer:* Well, of course, you have asked a very vast
> question that is filled with argument on both sides. You
> have raised the question of states rights versus Federal
> power; you have particularly brought up the question
> whether the Supreme Court is the last word we have in
> the interpretation of our Constitution.
>
> Now, this is what I say: there are adequate legal means
> of determining all of these factors. The Supreme Court
> has issued its own operational directives and delegated
> power to the district courts.
>
> I expect that we are going to make progress, and the
> Supreme Court itself said it does not expect revolutionary
> action suddenly executed.
>
> We will make progress, and I am not going to attempt
> to tell them how it is going to be done.

The President did not wish to discuss the matter further. But
more than one reporter wondered just what "legal means"
could be used to determine "whether the Supreme Court is
the last word we have in the interpretation of our Constitution." In Chicago, newsmen asked Stevenson whether he

agreed with the many southern congressmen who had signed
a "Manifesto" declaring that the Supreme Court had gone
beyond its powers in its integration decisions:

> I do not agree that the Supreme Court exceeded its proper
> authority on school segregation. I think rather that these
> rulings are correct interpretations of the Constitution and
> the conscience of the nation.

A foreigner might reasonably have wondered, on reading
these two statements, which man was, in fact, President of
the United States.

## IV

To support the desegregation decisions forthrightly and to
reject interposition proposals on the one hand, and proposals
for immediate drastic federal action on the other, was still
not enough for a candidate with Stevenson's sense of responsibility. He felt obliged to suggest what he would do if
he should win the office he sought. In New York, on February 27, he made a proposal and commitment which he
often reiterated in the weeks and months following. The
suggestion he made was calculated at once to turn a partisan campaign to the uses of national leadership, to commit
himself to a clear course of action if he should become President, and to leave open the way for Eisenhower to act without political embarrassment should the latter wish to adopt
that course himself. Indeed, Stevenson's proposal was well-
suited to Eisenhower's talents for working harmoniously with
people of divergent views:

> I for one have been very much disturbed by mounting
> tension in the South. In order to avoid any possibility of

further disorders or further damage to the nation's reputation abroad, I think the situation merits the prompt attention of the President.

The office of President of the United States has great
moral influence and great prestige and I think the time
has come when that influence should be used by calling
together white and Negro leaders from the areas concerned
in the South to explore ways and means of allaying these
rising tensions.

Such a conference would strengthen the hands of the
thoughtful and the responsible leaders of both races by
whom such conspicuous progress has been made in desegregation and in maintaining good relations with the
races. The prestige of the President could curb the tensions
in the South. It should be exerted before the situation gets
any more serious.

Though many leaders of both races endorsed Stevenson's
proposal, the President was slow to respond. In his press conference of March 14 he expressed qualified approval of the
idea, but said that he preferred to have Congress authorize
a bipartisan joint commission with power to subpoena and
compel testimony. "This decision," he said, "was made in
1954, and we are getting along now to where some real
investigative body ought to be watching it." But to investigate
violations was not action to avoid them. The next day Senator Richard Neuberger, Democrat, of Oregon, called in the
Senate for both a conference such as Stevenson had recommended and a commission such as the President preferred.
Representative Adam Powell, Harlem Negro, for the third
time urged the President to call a conference, and a leading
rabbi proposed to the President that he call the conference
on a nationwide basis, as had been done in such matters as

federal aid to education. Governor LeRoy Collins of Florida
wrote to the President suggesting a conference of southern
governors and attorneys general with the President. Eisenhower replied that if his counter proposal of a bipartisan
commission were not approved by the Congress, he would
consider other possibilities, including such a conference as
Collins suggested. On April 24 a White House aide again
wrote to Governor Collins to say that Eisenhower was still
weighing the suggestion of a conference. Finally, on May 23,
Eisenhower decided against holding a conference of southern
governors for fear of "inflaming racial feelings." Fifteen
months later, with racial feelings inflamed and federal para-
troops escorting Negro children to school, a southern governors' conference was held at the White House—too late.

Stevenson, for his part, went on campaigning. On March
28 he was back in Los Angeles, resolved that the distorted
impression of his views on civil rights, created by the un-
happy meeting there in the previous month, should be corrected. He had just been shocked at his defeat by Senator
Estes Kefauver in the Minnesota primary, and was much
concerned about the political effect his rival might be creating in the California primary by a less moderate stand on the
issue. Stevenson now clarified his position in firm language:

> For four years I've done my level best to unite the
> Democratic Party, not to tear it apart. And I propose to
> keep on thinking that the party's welfare is just as impor-
> tant as my own candidacy. . . .
> Another thing: I said in 1952 that my position would
> be just the same in all parts of the country. It was then.
> And it will be this time, too. And that position will not
> be changed to meet the opposition of a candidate who
> makes it sound in Illinois as though he opposed Federal

aid to segregated schools, in Florida as though he favors
it, and in Minnesota as though he had not made up his
mind. . . .

I want this to be a place where men and women and
children of every race, creed and color share alike the
opportunities that this great country has to offer.

Once more he took his stand on integration:

And that reminds us that eliminating segregation in the
schools of some of our sister states presents us today with
a national challenge to our maturity as a people. For my
part, like most Northerners, I feel that the Supreme Court
has decreed what our reason told us was inevitable and our
conscience told us was right. I feel strongly that whether
you agree with that decision or not, it is the law and should
be obeyed.

Here he laid bare the great difference between himself and
Eisenhower on the whole matter—a difference which future
events were to dramatize tragically. While Eisenhower, under
pressure of events, would recognize the desegregation decisions as the law, and act finally to uphold it, he would not
give the law his personal and positive support backed by his
popularity and prestige. Stevenson would endorse the law
first, then act to support it. With Eisenhower, partisans on
either side of the integration line could look the other way.
With Stevenson, opponents of integration, if they supported
him at all, would need to hold their opinions in private,
knowing that to support him was to support a man committed
to fulfilling in practice a conviction of conscience as well as
law.

Again Stevenson underscored his sense of personal responsibility for progress and reiterated his proposal for presidential action:

> It isn't enough, though, for any responsible public official or for any candidate for public office, just to say which side of this issue he is on, or how strongly he feels about it. The job that has to be done now is to find even in the conflicting counsel of those who disagree so violently the best course by which the Court's decision can be carried out.
>
> The Supreme Court has said what is to be done. The courts will determine when compliance will be expected. The question of how we will effect this transition in an orderly, peaceful way remains to be settled. This question is not going to settle itself. And the longer we drift the greater the danger—the danger from those who would violate the spirit of the Court decision by either lawless resistance or by undue provocation.
>
> I have suggested that the President should promptly bring together white and Negro leaders to search out the way to meet this problem as a united people. . . .
>
> . . . I believe that no man, North or South, has any greater present duty than to help find the way to unite this nation behind the right answer to this problem.

Some liberals were still not satisfied—could not be satisfied by such a position. But evidence mounted that within the Democratic party of California Stevenson was creating an expanding center of unity. Not the least significant aspect of his tremendous primary victory some two months later was the fact that he rolled up some of his widest margins in Negro wards of the big cities.

Once more in his primary campaign—this time before a

nationwide audience—Stevenson devoted a full length address to civil rights. At New York, on April 25, he again spoke of the responsibility of the President to lead the nation toward peaceful integration of the schools:

> The achievement of equal rights for all American citizens is the great unfinished business before the United States.
>
> This would be just as much the case had there been no Supreme Court decision on desegregation in the public schools.
>
> There remains, however, the abiding responsibility of the executive branch of the government to do its part in meeting this most fateful internal problem and the rising tensions that have followed in its train. The present administration, in my judgment, has failed to meet this responsibility; it has contributed nothing to the creation of an atmosphere in which this decision could be carried out in tranquility and order.
>
> The immense prestige and influence of the Presidency have been withheld from those who honestly seek to carry out the law in gathering storm and against rising resistance. Refusing to rise to this great moral and constitutional crisis, the administration has hardly even acknowledged its gravity.
>
> It is the sworn responsibility of the President to carry out the law of the land. And I would point out that this office is the one office of the democracy, apart from the courts, where the man who fills it represents all the people. Those in the Congress bear particular responsibilities to the citizens of the states they represent, and on what is in some ways a regional problem views are naturally divided. Not so of the Presidency. And where the nation is divided,

there is special demand on him to unite and lead the people toward the common goal.

There was understanding here of the political and moral problems faced by southern leaders of his party in Congress —an understanding aimed at avoiding an irrevocable split —and aimed, too, at consolidating his own leadership. But there was no concession:

> As President, if that were my privilege, I would work ceaselessly and with a sense of crucial urgency—with public officials, private groups, and educators—to meet this challenge in our life as a nation and this threat to our national good reputation.
>
> I would act in the knowledge that law and order is the Executive's responsibility; and I would and will act, too, I pray, in the conviction that to play politics with the Court's decision and the basic rights of citizens and human beings is wicked.

Stevenson now turned to another aspect of civil rights, and stated a position that was to have far-reaching consequences:

> It is, for me, another point of central principle that one of the most important guarantees of equality is the right conferred on all citizens by the Fifteenth Amendment. Political freedom underlies all other freedoms. Wherever any American citizens have been denied by intimidation and violence the right to vote, then the right to vote of all American citizens is imperiled.
>
> There are laws on the statute books giving the federal government authority to protect citizens' voting rights.

These laws should be enforced. And if they are inadequate, they should be strengthened. And I believe that all responsible citizens, south and north, would go along with a determination to assure the right to vote under the law.

Finally, Stevenson again challenged President Eisenhower to act:

But again I say that the responsibility and the opportunity of the President in matters such as these go beyond the execution of the law. For his office is the repository of moral authority as well as legal, and it is from our chosen leader that there must come to our hearts, our heads and our spirits the common impulse to honor in our daily lives the principles we proclaim to all the world.

In pointing specifically to the need for enforcing the right to vote, Stevenson articulated a growing national sentiment. The Eisenhower administration had given a number of indications of its awareness of this need. But the efforts of Senator Thomas Hennings, chairman of the Senate Subcommittee on Constitutional Liberties, to elicit specific legislative proposals from the Attorney General were inconclusive. Whether the President and his advisers wished to withhold their proposals until after the election, when they hoped to have a Republican Congress to receive credit for action on civil rights, or whether it was thought politically prudent for the President not to be committed until after his reelection, cannot be learned from the public record. But it is certain that the Eisenhower administration did not, in fact, offer a civil rights bill until the Eighty-fifth Congress, again Democratically controlled, convened in 1957.

V

With his victories in the Oregon, Florida, and California primaries Stevenson put to rout all his opponents for the nomination, whether they had taken "strong" or "weak" stands on civil rights—all, that is, except Governor Averell Harriman of New York, who had not entered the primary contests. Since Harriman rejected Stevenson's policy of moderation on civil rights in favor of more immediate and drastic federal action, it was necessary to wait for the Democratic Convention itself to learn whether Stevenson's views had succeeded in producing substantial unity in the party. While the evidence seemed convincing that he had done so, there was a day or two of suspense at the opening of the convention as former President Truman dramatically intervened to support Harriman. Truman's name was identified in the public mind with a stand for civil rights which had split the party in earlier years, and Harriman presented himself as a "Fair Deal" candidate on civil rights, as on other issues.

But when the platform committee reported out a moderate, Stevensonian plank on integration, voting, and other civil rights, Mr. Truman expressed himself as entirely satisfied and spoke from the floor on behalf of the plank. Thereafter Stevenson's "band wagon" was unstoppable. His nomination on the first ballot symbolized a greater degree of unity in the Democratic party than it had known for many years. Despite southern dissidence, the party was committed, like Stevenson, to recognizing the integration decisions and to action to enforce voting rights. The delegates themselves knew that their candidate would certainly speak more strongly than the platform, and act more positively if elected. Yet no splits occurred and even sorely disappointed delegates remained at their posts. Thus Stevenson's best contribution to the cause of civil rights was made in the act of winning the nomination. In doing so he showed that a measure of unity on civil rights could be achieved within the party and thus, on this issue, in the nation, and vindicated his views as those best calculated to make national progress. He could now accept his nomination as the unquestioned leader of the majority he had molded and, again as a partisan, speak for a growing national majority:

> Nor has it [the Democratic platform] evaded the current problems in the relations between the races who comprise America, problems which have so often tormented our national life.
>
> Of course there is disagreement in the Democratic party on desegregation. It could not be otherwise in the only party that must speak responsibly and responsively in both the North and the South. If all of us are not wholly satisfied with what we have said on this explosive subject, it is because we have spoken the only way a truly national party can.
>
> In substituting realism and persuasion for the extreme of force or nullification, our party has preserved its effectiveness, it has avoided a sectional crisis, and it has contributed to our national unity as only a national party could.
>
> As President it would be my purpose to press on in accordance with our platform toward the fuller freedom for all our citizens which is at once our party's pledge and the old American promise.

As the election campaign began, Eisenhower told his press conference on August 8 that he did not know whether the Republican platform would have a plank on integration and had given no thought to its wording. When asked if he had plans for enforcing the integration decision, he again asserted that this was a matter for the courts, not the administration. Again, on September 1 he avoided the issue of the Republican civil rights platform because, he said, "no plank can please all." Asked specifically whether he would endorse the integration decision, he declined—"it makes no difference whether or not I endorse it." At the Republican Convention a "strong" civil rights plank was revised after consultation with the White House. But in its final form the Republican plank was still slightly stronger than the Democratic.

On October 11 Eisenhower further explained his stand on the court decision by saying that he was sworn to uphold the Constitution, whether or not he agreed "with every single phrase" of it. The next day, on nationwide television, he said that the United States "won't be easy with its conscience" until everybody has equality and opportunity as "visualized by the Constitution," but did not mention the court decision. On October 14 Congressman Powell, a Democrat, announced that he was switching his vote to Eisenhower, one reason being that Eisenhower had agreed not to oppose the Powell Amendment forbidding federal aid to segregated schools. But the White House promptly denied there was any agreement with Powell. Again in 1956, Eisenhower said nothing in the South about civil rights.

Stevenson had spoken on civil rights so frequently and in so many places during the months of his primary campaign that he decided to give most of his attention, in this second contest with Eisenhower, to other issues, especially national defense and foreign policy. But even as the crisis in the Middle East mounted, threatening serious involvement of the United States, he felt it necessary to speak out once more on the gravest domestic issue since Appomattox. And again the circumstances were dramatic. In the event those circumstances were more filled with drama than anyone could have foretold, for the place he chose to speak was the public square in Little Rock, Arkansas:

> Change has brought great opportunities, and it will bring more. Yes, and change also brings grave problems, and these too are the concern of the national community, acting through the national government.
>
> There is today a critical division of national opinion regarding recognition of common rights of American citizens of different racial origins. This division is reflected in the Democratic Party—necessarily reflected because, unlike our opposition, ours is a national party which has its roots in every section of the country.
>
> I find reason for great encouragement in the fact that the Democratic Party has risen above this division. Here is the promise, the assurance, that the nation too will rise above this division.
>
> The Supreme Court of the United States has determined unanimously that the Constitution does not permit segregation in the schools. As you know, for I have made my position clear on this from the start, I believe that decision to be right.
>
> Some of you feel strongly to the contrary.
>
> But what is most important is that we agree that once the Supreme Court has decided this Constitutional question, we accept that decision as law-abiding citizens.
>
> Our common goal is the orderly accomplishment of the result decreed by the Court. I said long ago and I stand now squarely on the plain statement, adopted in the Democratic platform, that "We reject all proposals for the use of force to interfere with the orderly determination of these matters by the courts." The court's decree provides for the ways and means of putting into effect the

principle it sets forth. I am confident that this decision will be carried out in the manner prescribed by the courts. I have repeatedly expressed the belief, however, that the office of the Presidency should be used to bring together those of opposing views in this matter—to the end of creating a climate for peaceful acceptance of this decision.

The speech was applauded. A few days later Stevenson re-enforced the consistency of his policy by saying the same things at a rally in Harlem:

They [the Republicans] have even claimed credit for ending segregation in the District of Columbia—though the case which meant the end of segregation in many public places in the District was initiated at the time President Truman was in office, and while Mr. Eisenhower was a private citizen.

And finally, when the President was presented with an opportunity for great national leadership in this field, he was virtually silent. I am referring to the Supreme Court decision on desegregation in the public schools.

Surely the gravest problem we face here at home this year is this issue of civil rights. . . .

Yet despite the progress we have made, the achievement of equality of rights and opportunities for all American citizens is still the great unfinished business before the United States. The Supreme Court decision on desegregation in the public schools was an expression of our steady movement toward genuine equality for all before the law: it expressed in a new field the old principle that the American heritage of liberty and opportunity is not to be confined to men, women, and children of a single race, a single religion, or a single color.

Then he quoted directly from his Little Rock speech. Finally,

> I pray that all Americans, no matter what their feelings, will collaborate in working to sustain the Bill of Rights. No other course is consistent with our constitutional equality as Americans or with our human brotherhood as children of God.

## VI

At the polls in November the nonpartisan President scored an even greater popular victory than he had won in 1952. To the four southern states he had carried on the first attempt he added Louisiana and Kentucky, though, significantly, for the first time Stevenson carried the border state of Missouri, where school integration was quietly proceeding. The election fell at a moment when war was raging in the Middle East and the whole world was alarmed lest it should spread uncontrollably. The American people endorsed the hero-general, now more than ever carrying the image of a beloved father, for another term as national symbol. But their votes for a man who stood above politics could not resolve the issues of a divided people. Nor could the President provide leadership without alienating a significant element of his popular support. At any rate, he did not.

The second defeat of Adlai Stevenson was a paradox, since a people who had voted for an uncommitted President, perhaps largely because he *was* uncommitted, voted in their majority to repudiate the President's party and returned the Democrats to control of Congress with a greater majority than before.

The principal business of the new Congress was civil rights, and the administration at last proposed a full-scale program. But when the debate and political maneuvering were done and action taken, it was an agreement on voting rights, plus a federal commission, that was enacted into law. Five Southern Democratic Senators joined their colleagues from the North to pass the first civil rights statute since the day of Reconstruction. The law reflected the maximum of agreement on civil rights to be found in the nation. It was this measure of agreement that Stevenson had sensed and articulated in his campaign of 1956. His nomination and leadership thus far saved the Democratic party from a disastrous break-up and made the civil rights law possible. As a presidential candidate and as a private citizen he had been able to lead where Eisenhower could not or would not.

And still the President waited as the school crisis mounted. Despite repeated and urgent requests that he go into the South to speak of civil rights and call a conference of leaders, Eisenhower refused to act. In September, 1957, he and the nation reaped the bitter harvest in inaction. When the Governor of Arkansas called out the militia of his state to prevent the integration of the school in Little Rock, in defiance of federal court orders, Eisenhower was at last forced to act. His only recourse was to send in troops of the regular army and call the state militia into federal service. In the end the only possible measure was that use of force which both he and Stevenson had feared and deplored. The consequence of political detachment and disengagement was military engagement. When southern governors at last met with the President on October 1, 1957, it was to consider ways and means of withdrawing soldiers from a scene of strife. Whether a White House conference of southern leaders, held long before and led by a President patient but positive and firm, could have precluded the national disaster of

Little Rock, no one can say. But in perspective it can scarcely be argued that such a course would not have been worth the attempt.

After Little Rock there were signs even among Eisenhower's most devoted supporters that his indecision as a leader was beginning to be recognized. In October, 1957, *Life,* long known as a strong and consistent voice for the Eisenhower administration, editorialized in these words:

> . . . where the President is open to criticism is not for exceeding his authority but rather, that in using it, he left room for doubt as to whether he himself believes in the law that he is enforcing. In his speech to the nation he remarked, "Our personal opinions about the (desegregation) decision have no bearing on the matter of enforcement," just as, on earlier occasions, he has stated that his own opinion "makes no difference." Such attitudes, though technically correct and politically prudent, have left room for inference that the President equates the 14th Amendment with the 18th (Prohibition), a disagreeable thing which has to be enforced even though it may be unwise.

The lesson for responsible American citizens may well be that popularity born of detachment from vital issues cannot safely be confused with leadership, and that when divisions are allowed to drift into crisis for lack of a committed leader, extreme measures deplored by every one may be necessary to fill the vacuum left by wasted opportunities.

Adlai Stevenson's comment on Little Rock was brief:

> At this point the President had no choice. The combination of lawless violence and the Governor's irresponsible behavior have created a crisis which Arkansas is powerless

to meet. Federal force must in this situation be used to put down force. But this is only a temporary solution. We have suffered a national disaster and I hope the President will now mobilize the nation's conscience as he has mobilized its arms.

A few days later, when President Eisenhower at last announced the calling of a conference of southern governors at the White House, Stevenson was in Chapel Hill, North Carolina, to address a Citizens Committee for Better Schools. At a press conference he was asked to comment on the President's action. In a tone of sadness that seemed to flow out of the words, he had this to say:

It is no secret that almost two years ago I urged such meetings as President Eisenhower has arranged for next week with the Southern Governors. I wish he had conferred with them and leaders of both races long ago. If he had done so and his position had been clearly expressed I don't believe we would have suffered this national misfortune which has been so widely exploited by our enemies.

It is time now to bind up our wounds. I hope the soldiers can be quickly withdrawn from Little Rock, and that local authorities and law-abiding citizens everywhere will see to it that they are never needed again anywhere.

## CHAPTER IV

# Foreign Policy:

# From Korea to Quemoy

## I

WHEN Dwight Eisenhower and Adlai Stevenson first contested for the Presidency in 1952, armistice negotiations in Korea had been going on without success for many months and the war still raged. For both men, as for the nation and the world, Korea was the central, overriding issue. By accepting the nomination Stevenson, as Democratic candidate to succeed President Truman, had committed himself to defending the administration. Since the war was bitterly unpopular, he could not hope to gain political advantage by his role. But he could defend the war with conviction and hope to neutralize its effect on the election by patient explanation and forceful moral exhortation. Eisenhower, by deciding to criticize the administration, both for permitting the war at all and for its handling of strategy, and by suggesting that peace could be achieved the sooner by his election, responded directly to the wishes of the popular majority and built up an insurmountable advantage. But his very success with the issue brought with it imperative expectations for the future which could only limit his powers to lead the nation in foreign affairs. He was thereafter committed to a "tough" line with Communists everywhere, to avoiding any sign of "appeasement," and yet to reconciling opposites toward a world at peace.

In the earliest stages of the campaign, Eisenhower and other Republican speakers treated the Korean War chiefly in a minor key. Before his nomination the General had made a number of statements about Korea which were so close to the official position of the Truman administration as to yield political discomfort to him later in his role of nominee. But as the campaign moved along, there was a drumfire of criticism of Truman for "bungling," and there was constant repetition of the theme that Stevenson would only carry on a foreign policy that had failed. Stevenson, for his part, answered the criticism with earnest words in defense of the war and of the policy that had led to American involvement in it. At Louisville on September 27, for example, he summed up his views:

> And let's talk sense about what we have gained by our determination, our expenditures, and our valor in Korea.
>
> We have not merely said, we have proven, that communism can go no further unless it is willing to risk world war.
>
> We have proven to all the peoples of the Far East that communism is not the wave of the future, that it can be stopped.
>
> We have helped to save the peoples of Indo-China from communist conquest.
>
> We have smashed the threat to Japan through Korea and so have strengthened this friend and ally.
>
> We have discouraged the Chinese communists from striking at Formosa.
>
> We have mightily strengthened our defenses and all our defensive positions around the world.

We have trained and equipped a large army of South Koreans, who can assume a growing share of the defense of their country.

We have blocked the road to communist domination of the Far East and frustrated the creation of power which would have threatened the whole world.

We have asserted, and we shall maintain it, that whenever communist soldiers choose freedom after falling into our hands, they are free.

We have kept faith with our solemn obligations.

Eisenhower seems to have reached his decision to focus his campaign on the Korean War in the early days of October. At any rate, he began on October 2 to capture the popular imagination with slogans and proposals that Stevenson, from his defensive position, could not match. At Champaign, Illinois, Eisenhower told his audience:

There is no sense in the United Nations, with America bearing the brunt of the thing, being constantly compelled to man those front lines. That is a job for Koreans. We do not want Asia to feel that the white man of the West is his enemy. If there must be a war, let it be Asians against Asians, with our support on the side of freedom.

Whether the groundswell toward Eisenhower had or had not already reached victorious proportions, as the pre-election polls were suggesting, it is certain that these words—headlined "Let Asians Fight Asians"—gave the General full control of the issue from that moment.

A week later, at San Francisco, he reviewed the background of the war in American policy. In one of his most skillful speeches he turned Democratic insistence that he was himself in part responsible for that policy to effective political advantage:

> Today this bloody line [38th parallel in Korea] marks the "defense perimeter" of our country in that part of Asia. Yet scarcely more than two years ago the present Administration announced its political decision that the "defense perimeter" of America in that part of the world was quite a different line. That defense line did not touch Korea—or indeed any part of the mainland of Asia—but ran through islands well off the continental shore.
>
> Many an American family knows only too well how history has dealt with this policy decision of our Government. The Communists hastened to exploit it. And we Americans are still paying dearly.

But this "political decision" had not only led the nation into the bitter war, it had betrayed Eisenhower himself:

> I remember well, of course, that in 1947 the Joint Chiefs of Staff made a secret military appraisal of the strategic importance of Korea to our armies in the event of a general war in the future. I was Chief of Staff of the Army at the time. As always, the Joint Chiefs were careful to refrain from political judgments that were beyond their authority.
>
> But there were some things back there in 1947 that I didn't foresee would happen.
>
> First: I didn't foresee that—three years later—the Secretary of State would translate that strictly military appraisal for war conditions into a peacetime political decision.
>
> Secondly: I failed to anticipate that—three years later —the Secretary of State would make public this political decision to a potential enemy.

Third: I certainly failed to foresee that—five years later
—this military assessment of a possible war situation would
be used by a desperate Administration as the excuse for
the political decision which it took in exercise of its civilian
responsibilities entirely on its own initiative.

Yes. I failed to foresee that there would be such a lack
of courage and candor in high public office.

Thus Eisenhower placed himself above politics, where, as
the election proved, the people wanted him to be. As he drew
his distinction between "civilian" and "military," he did not
remind his listeners that all government decisions are "political"; that in the United States the military is not a thing set
apart but a service at the disposal of a civilian government
and the people it represents. It was skillful if sophistical politics. And it was followed by an appealing pledge, the more
effective because he was a military hero and above "politics":

Here's what I pledge you:
Without weakening the security of the free world I
pledge full dedication to the job of finding an intelligent
and honorable way to end the tragic toll of American
casualties in Korea.
Here one way is pointed. I shall never say as the present
Administration says: Because the problem is tough, the
problem can't be solved. . . .

That his pledge was in fact the policy of the government he
sought to defeat at the election, that the administration (and
Stevenson) had said that the tough problem *could* be solved,
made no difference. Here was a fresh and authoritative voice
which seemed to promise an end to frustration and suffering;
it was the voice and the words Americans wanted to hear.
At Los Angeles the next week, Stevenson undertook to

reply. He analyzed the various possibilities available to Eisenhower in trying to carry out his pledge. First, there was the possibility of withdrawing from Korea, as Senator Capehart and other Republican leaders were suggesting:

> What effect do you think such a policy would have on reckless men? What effect on the free peoples living in the long, dark shadow of the Kremlin?
> Well, I will tell you what I think. I think that Korea is not the last ambition of the Soviet rulers. Far from it. I think that weakness will never persuade the Soviet rulers to keep peace. Once they had us on the run they would undertake new acts of aggression somewhere else, and if we pulled out of Korea, all of South Asia would be uncovered and inviting, like Indo-China, where the French have fought so long, so valiantly, and so expensively.

A second choice, Stevenson went on, was "to let Koreans do the fighting." But this, so far as possible, was already being done:

> It has been common knowledge, I thought, for a long time that the United States has been training and equipping Korean forces and that these forces are taking on more and more of the burden. . . . We know, too, on the authority and in the words of General Van Fleet, that Korean soldiers "will never be able completely to replace American troops in Korea as long as there is an active front."

Finally, there was the possibility that "only Asians should fight Asians." Such a view, Stevenson maintained, "completely misses the significance" of the war:

That war takes place in Korea, but surely no one—and least of all the General—thinks that the only object of the attack was the small territory of Korea and the twenty million citizens of the Republic of Korea.

The attack was aimed at America and the whole free world—and that is why many nations have responded. . . . The Korean War is not a war that concerns just Koreans. It is our war, too, because—and there should be no mistake about this—world domination is the ultimate target of the communist rulers, and world domination includes us.

A fourth possibility—extending the war to China—may be passed over, since it was never a real issue between Eisenhower and Stevenson, though at one moment Eisenhower did say that he had always agreed with General MacArthur on bombing Chinese bases behind the Yalu River. Stevenson turned now to his own proposal:

Now my own views regarding our course of action in Korea offer no miracles. . . .

Our purpose is peace. We must continue to press in every conceivable way the negotiations in Korea. I am keenly aware of the fact that most people in this country feel that there just must be some way that these negotiations can be pressed to a conclusion. We must keep the United Nations military forces in Korea at the strength which, with Koreans, will be necessary to withstand the Communist forces. Otherwise we will lose the negotiations and we will also lose the chance of settlement in Korea which we have now almost won.

Stevenson's logic was firm and forthright, but Eisenhower and the people were now riding on a wave of emotion no logic could restrain. And that wave of feeling had its own logic, a perhaps inevitable logic which drove the General to an ultimate personalization of his Korean policy. The climax came at Detroit, on October 24:

> In this anxious autumn for America, one fact looms above all others in our people's mind. One tragedy challenges all men dedicated to the work of peace. One word shouts denial to those who foolishly pretend that ours is not a nation at war.
>
> This fact, this tragedy, this word is: Korea.
>
> A small country, Korea has been for more than two years the battleground for the costliest foreign war our nation has fought, excepting the two world wars. It has been the burial ground for 20,000 American dead. It has been another historic field of honor for the valor and tenacity of American soldiers.
>
> All these things it has been—and yet one thing more. It has been a symbol—a telling symbol—of the foreign policy of our nation.
>
> It has been a sign—a warning sign—of the way the Administration has conducted our world affairs.
>
> It has been a measure—a damning measure—of the quality of leadership we have been given.

This was speech-writing of a higher rhetorical order than Eisenhower had yet offered. And the opening paragraphs set the tone and tenor of the whole. The General quickly said that he was "not going to give you elaborate generalizations." He would deal, he said, with the "unvarnished truth":

What, then, are the plain facts?

> The biggest fact about the Korean War is this: It was never inevitable, it was never inescapable, no fantastic fiat

of history decreed that little South Korea—in the summer of 1950—would fatally tempt Communist aggressors as their easiest victim. No demonic destiny decreed that America had to be bled this way in order to keep South Korea free and to keep freedom itself self-respecting.

Since the war was not "inevitable," it followed that someone was directly responsible for it. There was, of course, the responsibility of the North Korean and Chinese Communists who had attacked. But Eisenhower was concerned with a responsibility that, in his view, lay behind the act of aggression:

There is a Korean War—and we are fighting it—for the simplest of reasons: Because free leadership failed to check and to turn back Communist ambition before it savagely attacked us. The Korean War—more perhaps than any other war in history—simply and swiftly followed the collapse of our political defenses. There is no other reason than this: We failed to read and to outwit the totalitarian mind.

If the "simplest of reasons" was too simple, a majority of Americans were too impatient to be critical.

The General went on to consider the history of American policy in the Far East from the time of the Wedemeyer report on China and Korea, in 1947, down to the outbreak of the war. He stressed again Secretary Acheson's delimitation of America's "defense perimeter" as running outside of Korea. He recalled that Acheson had placed responsibility for defense of South Korea in the United Nations, but this, said Eisenhower, was "cold comfort" to the nations outside the perimeter. As though in response to an invitation, he seemed to suggest, the "enemy struck." Once the aggression had begun, Eisenhower agreed, to fight "was the only way

to defend the idea of collective freedom against aggression."
The appeal of the government to its youth to fight for freedom in Korea

> was inescapable because there was now in the plight into
> which we had stumbled no other way to save honor and
> self-respect.

If he should be elected, what would his administration do
to repair the tragic blunder he deplored, to restore peace—
what would he himself do about it?

> My answer—candid and complete—is this: The first
> task of a new Administration will be to review and re-
> examine every course of action open to us with one goal
> in view: To bring the Korean war to an early and honor-
> able end. That is my pledge to the American people.
> For this task a wholly new Administration would be
> necessary. The reason for this is simple. The old Adminis-
> tration cannot be expected to repair what it failed to pre-
> vent.
> Where will a new Administration begin?
> It will begin with its President taking a simple, firm
> resolution. That resolution will be: To forego the diver-
> sions of politics and to concentrate on the job of ending
> the Korean War—until that job is honorably done.
> That job requires a personal trip to Korea.
> I shall make that trip. Only in that way could I learn
> how best to serve the American people in the cause of
> peace.
> I shall go to Korea.

Thus ended one of the landmark speeches in the history of
presidential campaigning. If, in the next few days, Eisen-

hower carefully qualified its meaning by limiting the reasonable expectations of his proposed journey to the scene of
war, the effect was nevertheless perfected by these words.
And it was superbly done. Invoking the common distrust of
"politics," he identified "politics" not only with the administration but with the tragedy of the war. In so doing he
absolved himself from political taint. Politics could not undo
the damage: nonpolitical statesmanship and dedicated resolve
could. But the overtone was even more persuasive: the statesman absolved from politics was still the military hero of the
Second World War, and it was a war which he now proposed
to settle. A political philosopher, summoning all his powers
of imagination, could hardly envision a more politically fortunate circumstance. Nor could an orator find better words
to express it.

The rest of the campaign of 1952 was anticlimax. In vain
could Stevenson decry the turning of the Korean tragedy to
political advantage. If he had earlier felt some confidence of
winning the Presidency, he must have begun to entertain
serious doubts now. Though he strove by firm insistence on
courage and patience to recover the ground Eisenhower had
won, his speeches sounded a plaintive note. At Brooklyn,
on October 31, he made a final effort. His eloquence rose
well toward the occasion, but could not reach it:

> "Let Asians fight Asians" is the authentic voice of a re-
> surgent isolationism. In 1939 the Republican Old Guard,
> faced with the menace of the Nazi world, was content to
> say, "Let Europeans fight Europeans," ignoring completely
> the fact that the menace of Nazism was a menace to
> Americans as much as to Frenchmen and Englishmen.
> What a curious remark for a man who led the crusade
> against Nazi tyranny.

Addressing Eisenhower's pledge to go to Korea, he spoke harshly—perhaps justly—but also plaintively:

> This new proposal is simplicity itself. "Elect me President," the General says, "and you can forget about Korea; I will go there personally." I don't think for a moment that the American people are taken in by a promise without a program. It is not enough to say, "I will fix it for you." The principle of blind leadership is alien to our tradition. And, unfortunately, the ghost writer who provided the proposal failed to give it content. The General was to go to Korea, but nobody indicated what he should do when he got there. The American people were quick to realize also that the conduct of a military campaign is the task of a field commander, whereas the making of peace requires negotiation with the central adversary—and in this case the central adversary is in Moscow, not in Korea.

A Stevenson supporter interrupted to shout, "Let President Stevenson send General Eisenhower to Korea!" But this was whistling in the dark. Indeed, as the nation did not then know, Eisenhower's sudden gambit of going to Korea was especially bitter for Stevenson, since he had himself planned, if elected, to go to Korea in order to familiarize himself with conditions there. But he had decided not to mention this plan lest he raise false hopes and lay himself open to criticism for making politics out of the misery of war.

Looking back at the speeches of both candidates on the Korean War, one discovers that Stevenson's have better withstood the test of time. This is nowhere exemplified more clearly than in his closing speech at Brooklyn. Toward the end he dealt prophetically with an issue Eisenhower had avoided:

Korea was a crucial test in the struggle between the free world and communism.

The question of the forcible return of prisoners of war is an essential part of that test. Fifty thousand prisoners have stated that they would rather kill themselves than return to their homeland. Many of those prisoners surrendered in response to our own appeals. They surrendered to escape the communist world. They surrendered because in their eyes the United States stood for freedom from slavery. . . .

This is the sole question remaining unresolved in the truce negotiations. Is this the question General Eisenhower intends to settle by going to Korea?

Stevenson then cited the recently expressed views of Republicans like Senator Capehart, who were prepared to yield to China on the prisoner issue:

Now I think the General should answer one question. In embracing the Republican Old Guard has he embraced their contention that we give up our moral position?

Though the nation would not follow him, the end of his speech was powerful.

I have the profoundest sympathy for every mother and father in the United States who is affected by this tragic war. No one is more determined than I to see that it is brought to a conclusion. But that conclusion must be honorable, for if we do not maintain our moral position we have lost everything—our young men will have died in vain. If we give up on this point, if we send these 50,000 prisoners to their death, we will no longer lead the coalition of the free world. . . .

With patience and restraint and with the building up of our strength the Communists will be compelled to yield, even as they yielded on the Berlin Airlift.

As of the moment we have a stalemate, and stalemates are abhorrent to Americans. But let us not deceive ourselves. A stalemate is better than surrender—and it is better than atomic war. And let us not forget that a stalemate exists for our enemy as well as for ourselves.

There is no greater cruelty, in my judgment, than the raising of false hopes—no greater arrogance than playing politics with peace and war. Rather than exploit human hopes and fears, rather than provide glib solutions and false assurances, I would gladly lose this Presidential election.

## II

After his victory Eisenhower did indeed go to Korea. He surely lifted the spirits of the weary and frustrated soldiers of democracy holding the lines of the United Nations, and he raised the hopes of the free world. But he could not personally have ended the war; and he did not. It took the death of Stalin and a new order in Russia and more than half another year of embittered negotiations by the same team of negotiators who had begun their work under President Truman, before an armistice could be concluded. When the fighting at last ceased, it was the United States, on behalf of the United Nations, which made the decisive concession. The prisoners of war, whose moral value to the cause of freedom Stevenson had emphasized, were not in the end turned over to the Communists; but the Communists were allowed to "interview" them, to make sure of their defection and if possible, persuade them to return. Some were "persuaded" to go home, and thus the armistice produced a moral blur.

In another year the armies of Communist China, freed from the lines in Korea, marched southward to stand behind their Communist fellows in Indo-China, as Stevenson had foretold, and half that country passed behind the curtain of iron. Three years after the armistice, when Eisenhower triumphantly campaigned a second time and Republican speakers praised him as a peacemaker, the lines in Korea remained one of the world's trouble spots; and no peace was in sight. On the anniversary of the Communist assault on South Korea, June 25, 1958, the *New York Times* editorialized:

> On this anniversary we pay tribute to the gallant Korean fighters for freedom, and to our own men who have also made their sacrifice. But we cannot be insensible to the fact that our pledges have not yet been redeemed. We have accepted a stalemate. We have not yet made a just and lasting—and free—peace.

But most fateful for Americans, and for the free world, were the consequences to Eisenhower of the role he had adopted on the war in Korea. Because his own decision and the popularity he won and held kept him somehow elevated above the plane of crucial issues, his scope of effective leadership was limited to positions commanding nearly unanimous support. At home he could not successfully speak for both the isolationist and internationalist wings of his own party, so that when he needed Republican support in the Senate his foreign policy was distorted by Republican disharmony. Abroad he had at once to speak as the implacable foe of communism and as the voice of peace, so that his moving performance at the Geneva summit conference of 1955 was condemned to raise hopes that could not be realized. The chief burden of making decisive foreign policy statements fell almost inevitably on the shoulders of others—notably Secretary Dulles and Ambassador Stassen. But when their efforts produced controversy at home or disapproval abroad, Eisenhower was often moved to qualify or repudiate what they had said, and so to diminish the stature of his subordinates and confuse the image of America. Thus the Eisenhower administration denounced the "neutralism" of India and other countries, but expressed sympathy with it; "unleashed" Chiang Kai-shek, but required him to evacuate the Tachens; threatened "massive retaliation," but reassured European allies that they would not become a battlefield; prophesied an "agonizing reappraisal" of American policy in Europe, but continued the European-aid program; threatened American intervention in Indo-China, but saw the French withdraw in defeat; alternately approved and disapproved Nasser's Egypt; spoke strongly for Israel but refused her arms or a specific guarantee; asked for ever stronger ties with Britain and France, but was at last compelled to side with the Soviet Union against them in the United Nations. Such was the road from Korea to Suez.

### III

Adlai Stevenson, as titular head of his party and spokesman for the Democratic opposition, felt most keenly his responsibility in foreign affairs. Opposition to a government's foreign policy, in his view, was a quite different thing from criticism on domestic matters, for hostile or uncommitted powers might seek to exploit criticism of foreign policy as lack of unity, or even as indicating serious tension. And allies, too, might be troubled. Yet foreign policy, as Walter Lippman has said, is the "shield of the republic," and while a democratic opposition must support what it considers right, it cannot ignore what it considers government mistakes, lest the shield be fatally lowered.

Stevenson, by 1952, had already a rich background of experience in international relations. In his youth he had traveled as a journalist over much of the world and had visited the Soviet Union to see for himself how totalitarian communism behaved. In the 1930's he was a leader among midwestern groups who hoped to strengthen the democracies of Europe against the challenge of Hitler. During the Second World War he headed a United States economic mission to Italy, to survey that chaotic nation and make recommendations for its rehabilitation. When the war ended he took part in the founding of the United Nations at San Francisco, and was later chief of the American delegation to the United Nations Preparatory Commission in London. Afterward he served on several American delegations to the General Assembly and participated, notably, in the decisions which led to the partition of Palestine and the establishment of Israel as an independent nation.

After the 1952 election and his retirement as governor of Illinois, Stevenson decided to advance still further his knowledge and understanding of foreign affairs. In his own words:

> Out of a job—thanks to the voters—I went to see for myself. Starting from San Francisco in March, 1953, with four companions I traveled for six months around the edges of the Communist empire through Asia, the Middle East and Western Europe. I talked to the Emperor of Japan, the Queen of England, the Pope and to all the kings, presidents, and prime ministers along my route. And I also talked to countless diplomats, politicians, journalists, students, soldiers, peasants, porters, and multitudes of new and warm-hearted friends. Everywhere I encountered an eagerness to talk and a candor of expression among officials that touched and astonished me—and has heavily taxed my discretion. And often the hospitality

made me wonder if my hosts were confused and thought I had been elected President in 1952!

But this journey was more than a politician's junket. The four companions he took with him were carefully chosen. In addition to William McC. Blair, Jr., his able personal assistant, they were Barry Bingham, eminent journalist and publisher of the *Louisville Courier-Journal,* Walter Johnson of the University of Chicago, a skilled historian of American diplomacy, and William Attwood, a talented reporter of *Look* magazine. These men made their own highly useful contributions to Stevenson's mission of inquiry by observing conditions with alert eyes and interviewing countless people whom Stevenson could not reach—including even underground opposition leaders in some countries. The information and opinions thus diversely gathered were afterwards pooled in a common stock for Stevenson's use. Upon his return he published a series of articles in *Look* magazine and made a "Traveler's Report" on nationwide television. This address was nonpartisan, but it revealed the temper and theme of the opposition criticism Stevenson later leveled at the Eisenhower administration. For example:

> There is little tradition of democracy in these new states of Asia, but independence, won at long last, is a passion, which partly accounts in some quarters for their opaque view of Communist China where to many Asians it appears that the foreigners have been thrown out and the ignominy of centuries erased by Asians. There is reverent admiration for the ideas of the American Revolution, the Bill of Rights, and the great utterances of human freedom. But they think they see contradictions in waves of conformity and fear here at home, and hypocrisy in our alliances with the colonial powers and professed devotion to freedom and self-determination.

Again:

> The ideological conflict in the world doesn't mean much to
> the masses. Anti-Communist preaching wins few hearts.
> They want to know what we are for, not just what we are
> against. And in nations like India, Indonesia, and Burma
> they don't accept the thesis that everyone has to choose
> sides, that they have to be for us or against us. Nor do I
> believe that we should press alliances on unwilling allies.
> After all, we had a long record of neutrality and non-
> involvement ourselves, and the important thing is that
> such nations keep their independence and don't join the
> hostile coalition.

Finally:

> Some of the misunderstandings may seem incredible to us,
> but it is well to try to see ourselves as others see us. Many
> think we are intemperate, inflexible, and frightened. And
> people who have lived in insecurity for centuries don't un-
> derstand how there can be insecurity and fear in America
> which has never been bombed or lived in thralldom. Also,
> like ourselves, proud nations resent any real or suspected
> interference in their domestic affairs. Nor can they recon-
> cile our exhortations about the peril with deep cuts in our
> defense budget. And everywhere people think they recog-
> nize the dominant mood of America in what is called "Mc-
> Carthyism," now a worldwide word. Inquisitions, purges,
> book-burning, repression, and fear have obscured the
> bright vision of the land of the free and the home of the
> brave.

Toward the end Stevenson uttered a prophetic warning, two
years before the summit meeting at Geneva:

In these circumstances we should press forward—not under any foolish illusion that one grand conference would
yield security, but rather with realistic recognition that the
foundations of stability must be laid, stone by stone, with
patient persistence.

Throughout he emphasized a theme which was always central
to his conception of foreign policy and which moved, with
the passing years, ever higher on the national and world
agenda—economic development of backward areas:

But in spite of all their doubts and difficulties I was impressed by the devotion of the leaders of Asia to the
democratic idea of government by consent rather than
force, and by the decisive manner in which so many of
the new countries of Asia have dealt with violent Communist insurrections and conspiracies. Their revolutions
have not produced Utopia and they are struggling with in-
finite difficulties to raise living standards and satisfy the
rising tide of expectations. They want rice and respect, and
they want to believe in wondrous America that sends
friendly, earnest people to help them, and that believes
in them, and the aspirations of all God's children for peace,
dignity, and freedom.

In the spring of 1954, Stevenson delivered a series of
lectures at Harvard University, afterward published as *Call
to Greatness*. Here he soberly traced the development in
modern history of the crisis troubling the twentieth century,
analyzed the perils to the free world, and characterized
America's "burden" of responsibility to meet them. But at
the outset of the lectures he displayed an understanding of the
difficulties in conducting foreign policy which always underlay his criticism of Eisenhower:

It is easy to state our ends, our goals, but it is hard to fit them to our means. Every day, for example, politicians, of which there are plenty, swear eternal devotion to the ends of peace and security. They always remind me of the elder Holmes' apostrophe to a katydid: "Thou say'st an undisputed thing in such a solumn way." And every day statesmen, of which there are few, must struggle with limited means to achieve these unlimited ends, both in fact and in understanding. For the nation's purposes always exceed its means, and it is finding a balance between means and ends that is the heart of foreign policy and that makes it such a speculative, uncertain business.

With such knowledge and understanding Stevenson undertook the responsibility of criticising Eisenhower's foreign policy.

IV

In the campaign of 1954 from Alaska to New York Stevenson spoke often of foreign affairs. While congressional candidates concentrated largely on domestic problems in their own constituencies, his position as titular leader of his party, yet not a candidate for any office, enabled him to address the larger questions of America's relations with the world. His attack on the administration's "bluff and bluster," as he called it, voiced a concern among the people for the reputation of the United States, and his realistic appraisal of the international situation effectively countered Republican campaign slogans about Eisenhower's peacemaking. But an off-year congressional election is seldom a referendum on foreign policy, and the Democratic victory was inconclusive evidence of Stevenson's influence.

More direct evidence of Stevenson's constructive leadership was presently forthcoming, as a worldwide war scare developed over the fate of Quemoy and Matsu, tiny islands snug to the mainland of China.

As the new Congress convened in January, 1955, the Senate had before it for approval a treaty of mutual defense between the United States and the Republic of China, which at that time included Formosa (Taiwan), the Pescadores Islands, and a number of small islands at varying distances between Formosa and the mainland of China. While the Senate deliberated, Communist China began a series of minor military actions against the offshore islands, to which Nationalist China replied in kind. Whether Communist China hoped simply to intimidate the United States and force her to reconsider before the treaty went into effect, or whether Communist China in fact intended to launch an attack on Formosa, there is no way of knowing for certain. In later years it became apparent that provocative Nationalist sallies onto the mainland from these offshore bases were spurred by the U.S. Central Intelligence Agency.* What was certain at the time was that military actions took place almost daily and a threat of war existed.

On January 18 Communist China occupied Yikiang Island, thus threatening the larger Tachen Islands. At his press conference the next day President Eisenhower was asked to comment on this development:

> No military authority that I know of has tried to rate these small islands that are now under attack, or indeed the Tachens themselves, as an essential part of the defenses of Formosa and of the Pescadores, to the defense of which we are committed by the treaty that is now before the Senate for approval.

* See Fred J. Cook, "The CIA," *Nation,* June 24, 1961.

The two islands, I believe, that have been under attack are not occupied by Chinese National Regulars. They have been occupied by irregulars or guerrillas.

Now, the Tachens themselves are a different proposition. They are occupied by a division of troops. They are of value—and there is no denying that—they are of value as an outpost, an additional point for observation. They are not a vital element, as we see it, in the defense of the islands.

Now, exactly what is going to be the development there, I cannot foresee, so I won't try to speculate on exactly what we should do in that area. We don't even know, I think at this moment: at least, I wasn't informed this morning of what the Generalissimo's personal intentions are with respect to that particular region.

As Communist attacks continued and the fate of the Tachens remained in doubt, Eisenhower sent a special message to Congress, asking for authority to defend Formosa and the Pescadores. The key to his request was in these words:

I do not suggest that the United States enlarge its defensive obligations beyond Formosa and the Pescadores as provided by the treaty now awaiting ratification. But, unhappily, the danger of armed attack directed against that area compels us to take into account closely related localities and actions which, under current conditions, might determine the failure or the success of such an attack. The authority that may be accorded by the Congress would be used only in situations which are recognizable as parts of, or definite preliminaries to, an attack against the main positions of Formosa and the Pescadores.

In a matter of days the Democratic Congress had passed, with only three dissenting votes in each house, a resolution embodying the President's request.

While neither he nor the congressional resolution mentioned the islands of Quemoy and Matsu, there was no doubt that these were the "closely related localities" Eisenhower had meant. To defend them as a matter of policy would mean a new departure for the United States, since these small islands within a few miles of the mainland had always been Chinese territory, unlike Formosa and the Pescadores, which had belonged to Japan until the end of the Second World War. Thus for the United States to defend Quemoy and Matsu against Communist attack, for whatever reason, would constitute interference in the Chinese civil war. For this reason many nations, including America's Western allies, were alarmed at the prospect. On the other hand, many American supporters of Chiang Kai-shek and his government on Formosa, among them Republican leaders in the Senate, were calling for defense of Quemoy and Matsu as essential to the security of Formosa and as free-world outposts. At the same time it was clear that the Generalissimo himself hoped for American intervention to defend them. For both military and political reasons the President refused to commit himself. Thus in his press conference of February 2:

*Question:* Is there any indication that Chiang Kai-shek wants a public statement on Quemoy and Matsu?

*Answer:* Well, there are constantly, of course, conversations between our representatives and the Chinese Nationalists, and not always do our views exactly coincide. But I think that in view of the delicacy of the whole situation, one that, in its main parts, is before the United Nations, that it is better to stand for the moment on what we have said, at least publicly and let it go at that, and say no more for the moment.

The *New York Times,* nevertheless, reported on February 6 that "the United States has given the Nationalists private assurance that it would help, under present conditions, to defend Quemoy and Matsu." This view appeared to prevail throughout the world.

Early in February it was decided to evacuate the Tachen Islands. The United States Seventh Fleet assisted in this operation. Since this was the same fleet which had been "withdrawn" a year before in line with the Eisenhower administration's policy of "taking the wraps off Chiang Kai-shek" so that he could attack the mainland (though no one seriously supposed he could), the clarity of American policy was once more obscured. However, in a speech before the Foreign Policy Association on February 16, Secretary Dulles revived fears that the United States would go to war over Quemoy and Matsu:

> It has been suggested that the Nationalist Chinese should . . . surrender to the Chinese Communists' coastal positions which the Communists need to stage their announced attack on Formosa.
>
> It is doubtful that this would serve either the cause of peace or the cause of freedom. . . .
>
> The United States has no commitment and no purpose to defend the coastal islands as such. I repeat, as such. The basic purpose is to assure that Formosa and the Pescadores will not be forcibly taken over by the Chinese Communists. . . .
>
> . . . The Chinese Communists have linked the coastal positions to the defense of Formosa. That is a fact which, as President Eisenhower said in his message to Congress

about Formosa—and I quote—"compels us to take into account closely related localities." Accordingly, we shall be alert to subsequent Chinese Communist actions, rejecting for ourselves any initiative of warlike deeds.

In Formosa on March 3, to sign the papers of the Formosa–United States defense treaty, Dulles made the same point:

Since . . . the Matsu and Quemoy Islands now in friendly hands have a relationship to the defense of Taiwan, such that the President may judge their protection to be appropriate in assuring the defense of Taiwan and the Pescadores, our consultation covered also these coastal positions of the Republic of China.

Back in the United States on March 8, Dulles addressed the nation on the meaning of his trip to Formosa and of the mutual defense treaty:

The political decision of what to defend has been taken. It is expressed in the treaty and also in the law whereby Congress has authorized the President to use the armed forces of the United States in the Formosa area. That decision is to defend Formosa and the Pescadores. However, the law permits a defense which will be flexible and not necessarily confined to a static defense of Formosa and the Pescadores. How to implement this flexible defense of Formosa the President, President Eisenhower, will decide in the light of his judgment as to the overall value of certain coastal positions to the defense of Formosa, and the cost of holding these positions.

In the wake of Dulles' speech there were almost daily reports of "off the record" dinners and news conferences at which members of the administration were said to reveal an American commitment to fight for Quemoy and Matsu. The President, in his press conference of March 31, avoided the issue entirely. In Congress there were stirrings of concern that the Formosa Resolution had gone too far. Senators Lehman of New York and Morse of Oregon moved in the Senate to bar the use of American troops in defense of Quemoy and Matsu. Their proposal, though ineffective, was symptomatic of growing public concern.

In the early days of April the war scare was at its height. Would Communist China attack the coastal islands? And if so, would the United States go to war? In these circumstances there was a growing demand for Stevenson, as national Democratic leader, to speak his views. Letters and telegrams from all over the nation poured in a steady stream into his Chicago law office. But it was a difficult situation for him. Whatever he might say would have important consequences, not least in his own party. As a titular leader only, and holding no national office himself, he could not speak for the Democratic majority in Congress had he agreed with its views on foreign policy, which he did not. The Senate Democrats, in particular, sharing responsibility for foreign affairs with the President, had fully supported the President. They had passed the Formosa Resolution speedily and with little criticism. Stevenson thought the resolution unnecessary and ill-considered, since it seemed to commit the Democratic majority of the Senate to supporting in advance almost any line the President might choose to follow. They had given Eisenhower a blank check. Indeed, it was already difficult for Senators disturbed by the tension over Quemoy and Matsu to speak out against the administration. Stevenson thus would risk a split in his party, or perhaps alienate himself from powerful Democratic friends. At the same time, a formal expression of his opinions would have a significant effect on public opinion at home and abroad, both because of the great respect he personally commanded and because he would be speaking for the Democratic party as a whole. And finally, his own sensitive understanding of the intricate problems of conducting foreign policy made him hesitate to intervene at all.

Waiting until immediate tensions had subsided a bit, Stevenson chose a moment when he could exert a maximum influence for patience and clear thinking. After consultation with such friends as Chester Bowles, a former Ambassador to India, and with advance notice to Speaker Sam Rayburn, Senate Majority Leader Lyndon Johnson and Senate Foreign Relations Committee Chairman Walter George, Stevenson on April 11 addressed the nation by radio from Chicago. He began by recalling that it was just ten years since the United Nations had been founded at San Francisco with a "charter of liberation for the peoples of the earth from the scourge of war and want." But tonight, he continued:

> despite the uneasy truces in Korea and Indo-China, our country once again confronts the iron face of war—war that may be unlike anything that man has seen since the creation of the world, for the weapons man has created can destroy not only his present but his future as well. With the invention of the hydrogen bomb and all the frightful spawn of fission and fusion, the human race has crossed one of the great watersheds of history, and mankind stands in new territory, in uncharted lands.
>
> The tragedy is that the possibility of war just now seems to hinge upon Quemoy and Matsu, small islands that lie almost as close to the coast of China as Staten Island does to New York—islands which, presumably, have been fortified by the Chinese Nationalists with our approval and assistance.

Striking at Republican division on the issue, he went on:

> we now face the bitter consequences of our government's
> Far Eastern policy once again: either another damaging
> and humiliating retreat, or else the hazard of war, un-
> leashed not by necessity, not by strategic judgment, not
> by the honor of allies or for the defense of frontiers, but
> by a policy based more on political difficulties here at
> home than the realities of our situation in Asia.

Since the decision rested on the President's personal judgment as to the intent of any Communist attack on the islands,
it was not "improper," Stevenson said, to ask him, despite
his "great military experience," whether any man "can read
the mind of an enemy within a few hours of such an attack."
"Is it wise," he asked, "to allow the dread question of modern
war to hinge upon a guess?" He outlined the consequences of
a decision to go to war in a series of questions the President
must consider:

> Are the offshore islands essential to the security of the
> U.S.? Are they, indeed, even essential to the defense of
> Formosa—which all Americans have been agreed upon
> since President Truman sent the Seventh Fleet there five
> years ago?
> Or is it, as the Secretary of Defense says, that the loss
> of Quemoy and Matsu would make no significant military
> difference?
> Can they be defended without resort to nuclear weapons?
> If not, while I know we now have the means to inciner-
> ate, to burn up, much of living China, are we prepared to
> use such weapons to defend islands so tenuously related
> to American security?

Finally, are we prepared to shock and alienate not alone our traditional allies but most of the major non-Communist powers of Asia by going to war over islands to which the United States has no color of claim and which are of questionable value to the defense of Formosa?

Are we, in short, prepared to face the prospect of war in the morass of China, possibly global war, standing almost alone in a sullen or hostile world?

The questions answered themselves, and the tone of his voice, as he spoke, left no doubt as to the answers Stevenson himself would give.

Most important of all the elements in the Quemoy and Matsu peril was the risk of losing allies who could not agree to a belligerent policy by the United States. Stevenson turned next to this problem:

I know some politicians tell us we don't need allies. Life would certainly be much simpler if that were so. But it is not so. We need allies because we have only 6 per cent of the world's population. We need them because the overseas air bases essential to our own security are on their territory. We need allies because they are the source of indispensable strategic materials. We need, above all, the moral strength that the solidarity of the world community alone can bring to our cause. Let us never underestimate the weight of moral opinion. It was a general, Napoleon, who wrote that: "In war, moral considerations are three-quarters of the battle."

Because the great coalition, the alliance of free nations, must continue to be the basis of American foreign policy, Stevenson now proposed that a fresh start be made in dealing with the Formosa troubles by issuing a request for the advice both of "our friends" and of the "uncommitted states."

> . . . Ask them all to join with us in an open declaration condemning the use of force in the Formosa Strait, and agreeing to stand with us in the defense of Formosa against any aggression, pending some final settlement of its status—by independence, neutralization, trusteeship, plebiscite, or whatever is wisest.

Such a declaration would place the burden of responsibility for war, if war should come, squarely on the Communists, and would re-unify the free world. In addition, Stevenson proposed that the United States should ask the General Assembly of the United Nations "to condemn any effort to alter the present status of Formosa by force." This policy would repair "one of the weaknesses of our position . . . that we have been making Formosa policy as we thought best, regardless of others."

Having made his own suggestions, Stevenson next turned to a piercing criticism of the Eisenhower administration. He called for an end to "making threats" which the government "is not prepared to back up." He would not "belittle some recent achievements in the foreign field," but there is a "yawning gap between what we say and what we do." He cited the example of Indo-China, when the Vice President had "talked of sending American soldiers to fight on the mainland of Asia." This talk had ended in nothing, while half of Vietnam was lost. President Eisenhower himself had furnished a sad example of "these winged words"—his

> announcement two years ago that he was unleashing Chiang Kai-shek, taking the wraps off him presumably for

an attack on the mainland to reconquer China. However, it was apparent to everyone else, if not to us, that such an invasion across a hundred miles of water by a small, over-age, underequipped army against perhaps the largest army and the largest nation on earth could not possibly succeed without all-out support from the United States.

Since it seemed incredible to sober, thoughtful people that the government of the United States could be bluffing on such a matter, the President's unleashing policy has caused widespread anxiety that we planned to support a major war with China which might involve the Soviet Union. Hence we find ourselves where we are today—on Quemoy and Matsu—alone.

As he reached his conclusion, Stevenson made an eloquent plea for patience and for unity, and for a positive attitude toward peace:

If the best hope for today's world is a kind of atomic bal-ance, the decisive battle in the struggle against aggression may be fought not on battlefields but in the minds of men, and the area of decision may well be out there among the uncommitted peoples of Asia and Africa who look and listen and who must, in the main, judge us by what we say and do.

He deplored "the rattling of the saber" and an American posture which "made to appear hard, belligerent, and care-less . . . those very qualities of humanity which, in fact, we value most."

As best we can, let us correct this distorted impression, for we will win no hearts and minds in the new Asia by utter-ing louder threats and brandishing bigger swords. The fact is that we have not created excess military strength. The fact is that compared to freedom's enemies we have created if anything too little; the trouble is that we have tried to cover our deficiencies with bold words and have thus obscured our peaceful purposes and our ultimate reliance on quiet firmness, rather than bluster and vacillation, on wisdom rather than warnings, on forbearance rather than dictation. . . .

Let this be the American mission in the Hydrogen Age. Let us stop slandering ourselves and appear before the world once again—as we really are—as friends, not as masters; as apostles of principle, not of power; in humility, not arrogance; as champions of peace, not as harbingers of war. For our strength lies, not alone in our proving grounds and our stockpiles, but in our ideals, our goals, and their universal appeal to all men who are struggling to breathe free.

Thus Stevenson cast his influence against the risk of war over Quemoy and Matsu, and for a positive approach to the world crisis. There is no doubt that, partisan though he was, he spoke for the great majority of Americans. The next day, April 12, as though he had never suggested military intervention in the islands, Secretary Dulles said that Stevenson's proposals "copied" those of the administration. "Mr. Stevenson," he said, "has in fact endorsed the administration's program in relation to Formosa." Whether the Secretary's words meant what they seemed to say or were merely politic, there is no doubt that national unity on the Formosa question followed Stevenson's speech. No more was said of going to the military defense of Quemoy and Matsu, and in the discussions of the General Assembly overwhelming sentiment was expressed against the use of force in the Formosa Straits.

There is no reason to suppose that President Eisenhower

ever personally wished to go to war over Quemoy and Matsu. And, of course, it was the Communists who preserved peace by refraining from further attack. But Eisenhower was under severe pressure from leaders of his own party and from Nationalist China. Stevenson's intervention on behalf of a peaceful solution provided Eisenhower with the unity of American opinion he required to resist these pressures. In the moment of crisis the image of Eisenhower as peacemaker seemed to waver, but it was fortified and secured by Stevenson's leadership.

The second war scare over Quemoy and Matsu, in the fall of 1958, came as a kind of sorry repetition of earlier events. But there was one important difference among otherwise similar circumstances. In the interval between 1955 and 1958 Chiang Kai-shek had moved masses of troops to the small islands and fortified them heavily. This he had done with the consent—grudging, perhaps, but actual—of the United States government. When on August 23 Chinese Communist shore batteries began heavy bombardment of the islands, with accompanying threats and demands of surrender, the Nationalists had much more at stake than had been the case three years before.

Secretary Dulles, on behalf of the administration, immediately invoked the Formosa Resolution, reminding that President Eisenhower had authority to defend the islands if he deemed them involved in an attack on Formosa. In a speech on September 26 Dulles pointed out that "they [Communist China] say, and the Soviet Union says, that this is the beginning of an effort to take Formosa." After comparing the Quemoy–Matsu situation to the Korean armistice line, he went on to declare American policy in these terms:

When all factors, moral and material, are taken into account, their defense may not be divisible. So the United

States is assisting the Chinese Nationalists logistically in their gallant and inspiring defense of these positions. And President Eisenhower has in relation to these islands made clear that United States forces may be used more actively if the Chinese Communists push further a military effort which they themselves proclaim has Formosa as its goal.

Thus the United States was once again brought to the brink of war over the tiny islands. But this time it was quickly apparent that no considerable popular support could be generated for the administration's Formosa policy. The 1958 congressional election campaign was in full swing and Adlai Stevenson, campaigning in California on October 2, voiced effectively what public opinion polls showed was the predominant sentiment in the nation:

> We should make it clear that the United States is not helplessly entangled with nationalist China, that it is still master of its destiny, free, responsible and resolute.
>
> We should make it clear that we will fight to defend Formosa. It was surrendered by the Japanese after the last war, and we have as much right to be there as anyone. It is defensible militarily and philosophically. We are obligated by treaty to defend it, and let there be no doubt that we intend to defend it. And if we have no business in Quemoy, the Chinese Communists have no business in Formosa.
>
> But the offshore islands are another matter. They have historically always belonged to China. Their military value is negligible. They are indefensible, except by United States involvement at the risk of major war. The fight for these islands which have always belonged to China is a continuation of the Chinese Civil War in which we should not intervene.

Of course the United States must fight if need be to prevent communist aggression in the Far East. And we have—in Korea! And we will do so again if need be; let there be no doubt that the will-to-resist clear aggression burns as brightly today as it did in Korea.

But to stand fast when you are right is wisdom; to stand fast when you are wrong is catastrophe. To make our stand on Formosa's independence and self-determination would command the support of the great weight of world opinion. But to make our stand on Quemoy may imperil support on more important matters.

These, as Stevenson pointed out, were precisely the views he had expressed in 1955. Now more than ever they were the views of the country. On October 5 Chairman Green of the Senate Foreign Relations Committee wrote to President Eisenhower a message identical in its substance. The President replied somewhat pettishly that he would "welcome the opinions and counsel of others. But in the last analysis such opinions cannot legally replace my own." He expressed confidence that should we be brought to war we would be supported by our allies and by public opinion. Yet even as he wrote his administration was in fact responding to the great public pressures against involvement over Quemoy and Matsu. Chiang was persuaded to announce that he would not use force to return to the mainland; the garrisons on the islands were, at United States instigation, being reduced; and the tension was thus being relieved. On October 14, in response to a direct question at his press conference, Secretary Dulles glossed over this action in these words:

I have made no secret of the fact that over the past the United States has been inclined to feel that the troops there were excessive for the needs of the situation, and that view we still hold. . . .

We have no plans whatsoever for urging him [Chiang] to do that [reduce garrisons], although no doubt there are discussions that are going on over there probably at the present time between Secretary McElroy and others as to the most useful disposition of the forces of the Republic of China.

As in 1955, the incident came to an end when the Chinese Communists found it no longer useful to their strategy. But the startling parallel to the events of 1955 and the clear response of American opinion showed how far the nation had tended to unite upon views for years advocated by Stevenson, while Eisenhower's leadership had continually lagged behind his electoral success.

## V

After the 1955 furor over Quemoy and Matsu died down there was a moment of worldwide relaxation, climaxed by the Geneva summit conference in July of the same year. But scarcely had the heads of government returned to their capitals when a new crisis began to develop—a crisis destined to end in a sudden flash of war and a serious rupture of the Western Alliance at the very moment of the 1956 American election.

The origins of Israeli–Arab animosities lie in the remote reaches of history. But their immediate antecedents date from 1947, when the United Nations arranged a partition of Palestine and recognized Israel as an independent nation. The unwillingness of the Arab states of the Middle East to recognize the partition and the independence of Israel produced war in 1948 and thereafter a continuing obstacle to the United Nations in its efforts to maintain a truce. The Truman administration had favored the partition and received bipartisan support of its policy in Congress. And in 1950 Great Britain, France, and the United States made a Tripartite Declaration of their intention to support the independence of Israel and to oppose forcible boundary changes. At the same time, unsuccessful efforts were made to bring Israel and her Arab neighbors to agreement on resettlement of Arab refugees from Palestine. Border raids and clashes continued to erupt, and any moment of quiet in the Middle East tended to suggest the brewing of a new storm.

Pacification in the Middle East was a policy dictated by the immediate interests of the big Western powers. Great Britain and France were heavily dependent on the Suez Canal for cheap transportation of their commerce, and both, despite their withdrawal from the area as colonial powers, still had substantial investments there. American interests were complex but imperative. American companies were producing vast quantities of oil from Middle Eastern fields, where the largest part of the world's proved reserves is situated, and the oil itself was of vital importance not only to Europe but also to American military defense. Israel had no significant oil resources, so that effective American oil policy suggested improvement of relations with Arab states. But Israel's strongest allies, as has often been said, were the American Jews with their dedication to the integrity of the new homeland of their people. For all Americans, in addition, there was a special interest in Israel, since that new nation was an experiment in democracy, one of few such experiments in a part of the world still chiefly feudal. Thus interest and principle combined to suggest a Western policy of peace and settlement. The Soviet Union, on the contrary, could be expected to encourage tensions whenever it could do so without serious risk to its own interests. To penetrate the Middle East had, indeed, been a Russian objective for centuries.

Early in his term President Eisenhower (and Secretary Dulles) took the position that the removal of British troops from their base on the Suez Canal would reassure Egypt and other Arab states that Western intentions were peaceful. The British government reluctantly agreed to the suggestion and began troop withdrawals on October 19, 1954. As the British moved out, the Egyptians moved in; the exchange was completed on June 13, 1956, when the last British detachments departed. The Anglo–Egyptian Treaty was revised to permit the re-entry of the British in case of war, though the treaty could hardly envision a war between Britain and Egypt! Meanwhile, American arms were sent to Iraq as a defensive measure against possible Soviet threats from the North. And when President Nasser of Egypt announced his plans for a giant dam to be built at Aswan on the Nile, the United States indicated her interest in helping to finance the project. In the United Nations the American position was for the most part impartial regarding the almost constant charges and countercharges of Israel and her neighbors that borders were being violated in defiance of the armistice.

While some Democrats criticized the Eisenhower policy in the Middle East as "pro-Arab," or too narrowly devoted to the profits of the oil companies, Stevenson gave general support to the government. He had taken part, in the United Nations, in the establishment of the state of Israel, had visited there and numbered Chaim Weizmann and David Ben-Gurion among his friends, yet he recognized America's interest in Middle Eastern settlement and was convinced that it would in the long run benefit Israel to have the Arab states oriented to Western democracy. On his world tour of 1953

*Associated Press Photo*

"Let's talk sense to the American people."

*Above*. Running mates, 1952: John Sparkman of Alabama and Adlai E. Stevenson of Illinois. *Courtesy Reni Photos*

*Right*. Running mates, 1956: Estes Kefauver of Tennessee and Adlai E. Stevenson. *Courtesy Life-Magnum*

Whistle stop, 1952 campaign. With Stevenson are his aunt, Miss Letitia Stevenson, and his sister, Mrs. Ernest Ives.

Airport rally, 1956 campaign. At left of Stevenson is Tennessee Governor Frank Clement; at right is Senator Estes Kefauver.

*Courtesy Magnum Photos*

Campaign rally, 1956.

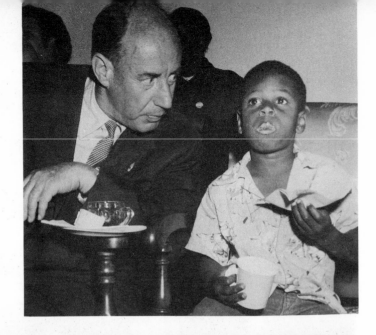

*Above.* With campaigners Mrs. Franklin D. Roosevelt and Senator Richard Neuberger, 1956. *Courtesy Oscar & Associates*

*Opposite page, top.* Two campaigners, 1956. *Courtesy Chateau Studio*

*Opposite page, bottom.* Writing a campaign speech in mid-air, 1956. *Courtesy Magnum Photos*

"How Did You Say Their Election Came Out?"

HERBLOCK
©1957 THE WASHINGTON POST CO.

*Above.* Stevenson in the Eisenhower administration, 1957. From *Herblock's Special For Today* (Simon & Schuster, 1958)

*Opposite page, top.* Stevenson's first appearance on the floor of the Democratic National Convention, 1960. *Courtesy Betty Furness*

*Opposite page, bottom.* Grass-roots sentiment, 1960. *United Press International Photo*

*Above.* With Premier David Ben Gurion in Israel, 1953. *Wide World Photo*

*Opposite page.* "A funny thing happened to me on the way to the White House." *Courtesy Magnum Photos*

With Dr. Albert Schweitzer in Lambarene, French Equatorial Africa, 1957.

"It would be a pity if our expectations were too high

*Courtesy Magnum Photos*

"The Russian people don't want war any more than we do." Soviet journey, 1958.

Soviet state farm in the Virgin Lands, 1958.

*Courtesy Magnum Photos*

With President Alberto Lheras in Bogota, Colombia, 1960.

In the slums of Rio de Janeiro, Brazil, 1960.

"We shall create with you working partnerships, based on mutual respect and understanding." Machu Picchu, Peru, 1960.

he had observed the elements of the Arab–Israeli problem at first hand and had talked at great length with the leaders on both sides of the uneasy armistice lines. In his *Look* articles, especially that of August 11, 1953, he had made a careful analysis of the difficulties besetting the whole area and deplored the lack of a firm and thoughtful American policy for pacification and development. His hope was for settlement through negotiation and integrated economic development. Stevenson expressed this view succinctly in an address to the American Committee for the Weizmann Institute at New York on December 2, 1954:

> As a friend of Israel since its inception, permit me to counsel patience and a broad perspective. In all fairness, one must see the Israel–Arab conflict in the context of the whole Middle East and Far East security program. One must, I think, accept the over-all assumption that an Arab world with a friendly orientation to the West is better for Israel than an Arab world with a friendly orientation in the other direction. One cannot, in good faith, take issue with the striving of our officials and other Western nations to improve relations, cooperation, and confidence in the Arab world. This would be a major goal of any administration in Washington.

It was Communist intervention in the Middle East that suddenly deflated the "spirit of Geneva" in September, 1955. The Egyptians announced a barter deal with Czechoslovakia —cotton for arms. As the arms shipments began to arrive in Egypt, it became clear to everyone that it was a very large-scale deal. With the arms went Communist technicians to teach Egyptians how to use them, technicians not only from satellite states but from the Soviet Union itself. Israel immediately appealed once more to the West, especially to the

United States, for help. It is perhaps safe to say that it was
the Communist arms to Egypt and the Israeli appeal for support which precipitated the Middle Eastern crisis of 1956.

When the Israeli government asked to buy fifty million
dollars worth of arms in the fall of 1955, such a request for
military assistance to Israel was not a novel problem to
President Eisenhower. At a press conference almost two
years before he had been asked whether

> . . . in view of the decision to grant military assistance
> to Iraq, has the Administration considered similar assist-
> ance to Israel?
>
> *Answer:* He had forgotten for the moment what was
> the state of our negotiations with Israel. He knew that we
> had rendered them economic assistance.
>
> Now we were not rendering anyone assistance to start
> a war or to indulge in conflict with others of our friends.
>
> We were—when we gave military assistance, that was
> for the common purpose of opposing communism, so if
> we did, and when we did, give military assistance to any
> region or any nation in that region, it was not for the pur-
> pose of assisting them in any local war of any kind.

The President thus measured the question of arms for Israel
against his broad policy of military alliances against communism. The clear implication was that arms to Israel would
not serve the chief purposes of his foreign policy, since such
arms would principally affect the balance of the issue between Israel and the Arab states, in which communism was
not thought to be involved.

When communism did become involved, through Soviet
penetration into Egypt in the fall of 1955, Eisenhower shifted
the grounds of his hesitation to send arms to Israel. On
November 16 at his press conference he laid down a policy

which characterized his administration as it prolonged hesitation and uncertainty for many months:

> A threat to peace in the Near East is a threat to world peace. As I said the other day, while we continue willing to consider requests for arms needed for legitimate self-defense, we do not intend to contribute to an arms competition in the Near East. We will continue to be guided by the policies of the tripartite declaration of May 25, 1950. We believe this policy best promotes the interest and security of the peoples of the area.

In a letter to Republican senators, released to the press on February 6, 1956, Secretary Dulles spelled out the administration's policy in greater detail:

> Let me say that the foreign policy of the United States embraces the preservation of Israel. It also embraces the principle of maintaining our friendship with Israel and the Arab states.
>
> The Government of Israel, feeling that its peaceful existence is threatened by large amounts of arms now made available to certain Arab countries by the Soviet bloc, desires to purchase from the United States and other countries additional armament to balance what it considers to be the increased threat against it.
>
> The United States recognizes the current developments could create a disparity in armed forces between Israel and its Arab neighbors. However, we are not convinced that that disparity can be adequately offset by additional purchases of arms by the state of Israel. Israel has a population of under 2,000,000 whereas the Arab population amounts to tens of millions, and they apparently have been offered access to huge stores of Soviet bloc material. Under

this circumstance, the security of Israel can perhaps better be assured by means other than an arms race.

Shortly thereafter, the issue was complicated by the disclosure that the United States had authorized a shipment of tanks and some other military equipment to Saudi Arabia. Eisenhower was asked at his press conference to explain why arms were still withheld from Israel in view of these shipments to an Arab state:

> Oh, well, now let's take that with a little bit of a grain of salt.
>
> Back in June or July of '55, it was agreed that a few light tanks and some auxiliary types of equipment could go to Saudi Arabia; and the only thing that was sent out there was material that had already been bought and paid for and export licenses issued a long time ago.
>
> Now the great thing the United States is trying to avoid is the initiation of an arms race in that region and because of that we have constantly restated our position that we believe that the United Nations should take urgent and early action on this matter, that both sides in the controversy should agree to abide by the United Nations' advice and armistice terms and avoid initiation—initiating incidents so we could get peace started.
>
> We do not believe that it is possible to assure peace in that area merely by rushing some arms to a nation that, at the most, can absorb only that amount that 1,700,000 can absorb; whereas, on the other side, there are some 40,000,000 people.

While Eisenhower could belittle the tank shipments to Saudi Arabia, the government of Israel could not. Surrounded by hostile neighbors who adamantly refused to recognize her independence and continued their pledges to destroy her, Israel could not reasonably be expected to see in the shipments anything but a willingness of the United States to let matters drift toward disaster. Talk of a "preventive war" was again heard among Israelis, though not by responsible officials. New representations for arms were made in Washington, but President Eisenhower's response, in a letter to President Ben-Zvi, was an appeal for "patience, mutual confidence and goodwill." On April 2 the White House press secretary announced that President Eisenhower and Secretary Dulles had decided to keep the Israeli request for arms "on ice." Thus, despite Communist intervention in the Middle East, the Eisenhower policy was not altered and matters did indeed drift.

Against a background of rapidly rising tensions in the Middle East and the successful Russian penetration there after more than two hundred years of failure, President Eisenhower addressed the annual convention of the American Society of Newspaper Editors on April 21, 1956, with these opening words:

> When I last appeared before this body, almost exactly three years ago, stories from battlefields and fighting fronts crowded the front pages of our press. Human freedom was under direct assault in important sectors by the disciples of communist dictatorship. Violence and aggression were brutal facts for millions of human beings. Fear of global war, or a nuclear holocaust, darkened the future. To many, the chance for a just and enduring peace seemed lost—hopeless.
>
> Today, three years later, we have reason for cautious hope that a new, a fruitful, a peaceful era for mankind can emerge from a haunted decade. The world breathes a little more easily today.

Now the prudent man will not delude himself that his hope for peace guarantees the realization of peace. Even with genuine goodwill, time and effort will be needed to correct the injustices, to cure the dangerous sores that plague the earth today. And the future alone can show whether the Communists really want to move toward a just and stable peace.

Yet not for many years has there been such promise that patient, imaginative, enterprising effort could gradually be rewarded in steady decrease in the dread of war; in an economic surge that will raise the living standards of all the world; in growing confidence that liberty and justice will one day overcome statism; in the better understanding among all peoples that is the essential prelude to true peace.

In his prepared speech there followed no mention at all of the Middle East. Afterward he spoke to the editors informally and at some length. But his only reference to the Middle East was to include British withdrawal from Suez among the achievements of his administration:

The difficulty in Egypt between our British friends and our Egyptian friends over the big base was finally settled.

## VI

When Eisenhower temporized on the sending of arms to Israel after the Czech–Egyptian arms deal in the fall of 1955, Stevenson withdrew all support of the Eisenhower policy in the Middle East. He had already regretted Eisenhower's failure to state firmly American opposition to change by force. Now the arms imbalance, he thought, was a serious threat to the integrity of Israel and presaged a renewal of hostilities between Israel and her Arab neighbors. He had declared his candidacy for the Democratic nomination and felt it necessary now to propose his own alternative course. In his address at the Woodrow Wilson Centennial in Charlottesville, Virginia, on November 11, he stated his position in positive terms:

> It is interesting and relevant to recall that Wilson believed in encouraging Jewish settlement in Palestine and took an active part in making the Balfour Declaration a vital part of the Palestine mandate under the League of Nations. Since then the state of Israel has become a fact, and unhappily, so also has the bitter hostility of its Arab neighbors. For five years violence along the armistice lines has been mounting. Unless these clashes cease there is danger of all-out war developing while we debate which side was the aggressor.

Under these circumstances, he went on:

> A major effort of statesmanship is required if we are to avert a political disaster in this troubled area. We have shown little initiative within or outside the United Nations in devising measures to prevent these border clashes. After years of experience it would seem evident that the only way to avoid bloodshed and violence along the border is to keep the troops of these antagonists apart. And I wonder if United Nations guards could not undertake patrol duties in the areas of tension and collision. Certainly both sides would respect United Nations patrols where they do not trust each other.

He then turned to the problem of an arms race:

> In this country we have been dismayed by the arms deal
> between Egypt and Russia. Of course there should be an
> equitable balance of armed strength so that neither side
> feels that it lives by the grace of its none-too-kindly neigh-
> bor. We must help, if need be, to counteract any Soviet
> attempt to upset such a balance, and we must make it
> emphatically clear that the status quo shall not be changed
> by force. But we do not want to see an arms race in this
> area where the principles of Woodrow Wilson's fourteen
> points once shone like a lighthouse after centuries of dark
> oppression.

Next, he warned the Arab states of Russia's long ambition
to penetrate their lands:

> And we trust our friends there have neither overlooked
> the fate of nations which have listened to the siren songs
> of Moscow, nor forgotten that the Soviet Foreign Minister
> told the Nazi Foreign Minister in 1940 that one of the
> conditions of a Nazi-Soviet agreement was that the Per-
> sian Gulf was to be a sphere of Soviet influence.

Finally, he said:

> We applaud the peaceful efforts of the Secretary General
> of the United Nations and we must bestir ourselves to help
> create conditions which will work toward peace, not
> conflict, in the troubled area. The United States does not
> choose sides when it chooses peace.

Perhaps the best comment that can be made on the policy
Stevenson advocated in this address was that he never after-

ward saw good reason to modify it. Developing circumstances tended emphatically to confirm his views. Nevertheless, his proposals could not be widely popular and, as he presently discovered, involved serious risks to his candidacy. To advocate the sending of American arms to Israel to restore the balance would be well received in Israel and among American friends of Israel, but it could not make friends among the Arabs or among supporters of Eisenhower's hands-off policy. At the same time, the crucial proposal of an armed United Nations patrol of the borders, to maintain peace by force, did not find favor with either Israel or the Arabs. Since both sides claimed innocence of aggressive intent, to welcome such military policing would be tantamount, in their view, to admitting responsibility for armistice violations.

Stevenson soon found himself the object of intense and conflicting political pressures within his own party and from independents interested in his candidacy. A battle "for the candidate's mind" developed on the Middle Eastern issue quite similar to the simultaneous "battle for his mind" on civil rights. And Stevenson dealt with it in the same manner as he did the civil-rights problem—he simply stated his consistent view in more positive terms. To avoid any misunderstanding he chose the occasion of a message to the American Jewish Committee, January 23, 1956, to re-emphasize what he had said at Charlottesville:

It seems to me that America's policy must be directed positively, vigorously and quickly to arresting the frightening tensions in the Middle East, not only for the sake of Israel but its Arab neighbors as well. Had it not been for Britain, France and the United States, these states of the Middle East would not exist as separate and independent nations. Surely we are entitled to see that independent nations for which we have made such sacrifices shall not be destroyed, let alone by one another.

A first step, as I have said before, should be the restraining effect of an equitable balance of armed strength between Israel and her neighbors. And I think that the security of all of these states should be guaranteed by the United States, and also by France and Britain who joined with us in the Tri-partite declaration against change by force. I have suggested, too, that one way to make such a guarantee effective would be to keep Israeli and Arab forces apart by substituting United Nations patrols in the areas of tension and collision on the borders.

These proposals, as he had said at Charlottesville, were intended for the short run. The long run required something more creative from the United States and a genuine effort by the hostile states of the Middle East:

. . . I think this country should reassert its fundamental friendship for both Israel and the Arab states, and its desire to help both develop economically, even as Israel is now developing. Once the Arab states recognize and accept the  permanence of Israel, then the real community of self-interest in the Middle East would become apparent and we can help solve the border adjustments, refugee re-settlement and other obstacles to realization of the great advantages to both sides of peace and economic progress in the Middle East.

His closing note was doubly prophetic, emphasizing both the Soviet threat and the necessity for the United States to act in concert with her great allies:

To this end and to frustrate the Soviet design to divide and dominate the Middle East, the United States must press on with a new and urgent resolve, in unison with Britain and France, whose interest in the integrity of Israel and the peace and prosperity of the whole Middle East is identical with ours.

There is no time to lose, for the issue in the Middle East today is not just the preservation of Israel but all the values of the Christian, Hebrew and Islamic cultures.

In his address to the American Society of Newspaper Editors on April 21, a few hours before President Eisenhower spoke to the same audience in such optimistic terms of the prospect for peace, Stevenson asked:

> Is the moral position of the United States clear and unambiguous and worthy of us and our real aims? Is the image of the United States one that inspires confidence and cooperation?
>
> Do you think we are winning or losing ground in the competition with the Communist world?

Stevenson—and the editors, after hearing both Stevenson and Eisenhower—thought the United States was losing. Stevenson was not content, however, simply to review the record. Among the positive proposals he offered was an important addition to his policy for the Middle East—an economic development program for the whole area, led by the United States and conducted through the United Nations. "Let me just suggest," he said, "that a coordinated attack on poverty in the Middle East might well be a profitable field for a United Nations economic program such as I have suggested."

In summary, Stevenson both as candidate for the Democratic nomination and later as candidate for President, offered to the voters a policy for the Middle East of four principal elements: arms to Israel to restore the balance that was being upset by Communist arms to Egypt; a direct United States–British–French guarantee of Israel's security; United Nations armed patrols on the borders to keep peace; a long-range program of economic development emphasizing the common interests of the Arab states and Israel, not their differences. It was a policy at once moderate and constructive, but poorly calculated to attract either votes or campaign contributions from interested minorities in the United States. Like his policy on civil rights it marked him as a prophet, an analyst, and a leader, if not as a winner.

## VII

While President Eisenhower hesitated, temporized, and finally refused arms to Israel, his administration took no positive steps of leadership to end the border clashes in the Middle East. If Stevenson's proposal of sending United Nations troops to patrol the borders was considered at all, it was not accepted. There is, indeed, some evidence that it may have been considered at a conference in Washington between British Prime Minister Eden and President Eisenhower. In their joint statement, issued in February, 1956, they said:

> We express our full support for the efforts of General Burns, head of the United Nations Truce Supervisory Organization, to maintain peace on the borders. We would favorably consider recommendations for any necessary enlargement of his organization and improvement of its capabilities.

The U.N. truce team was charged, of course, with observing the borders and reporting to the Security Council when violations occurred. If the Eisenhower–Eden statement meant that they were prepared to convert the truce team into an armed border patrol, which is certainly *not* clear, it *is* clear that nothing was done about it. Nor was any American leadership offered, in the United Nations or elsewhere, for a program of economic development in the Middle East. When he was asked at his press conference of March 21 what American program for the Middle East would be offered in the United Nations, Eisenhower answered, "The details of our plan will be published at the proper time. They are not out yet." And they never did come "out."

Meanwhile President Nasser of Egypt was pressing both his new-found advantage with the Soviet Union and his requests for money from the West to build the Aswan Dam. On July 19 Secretary Dulles announced that the offer of a credit to Egypt of $56,000,000 to commence construction of the dam had been withdrawn. Egypt, he indicated, had shown herself to be unstable by the heavy commitments of her cotton crop for Communist arms. One week later, in obvious retaliation, Nasser seized the Suez Canal. He proposed to pay a fair price to European stockholders and promised to give access to the canal to all its regular users. But the seizure gave Egypt control of tolls, and power to block the canal in case of war or to deny its use for political reasons. This seemed an intolerable situation to Britain and France who, like all of Western Europe, were dependent on Middle Eastern oil. It seemed likewise intolerable to Israel, since Egyptian control threatened permanent exclusion of her shipping from the canal. Secretary Dulles intervened to mediate between the Western powers and Egypt and, during the summer and early fall months, improvised a series of proposals which evoked little enthusiasm on the allied side and none on the Egyptian. Tense and hasty conferences were held on both sides of the Atlantic. But Egypt refused all suggestions, and tension rapidly mounted.

At various brief moments in September and October it appeared that some temporary accommodation between the West and Nasser might be reached. At the last such moment, October 12, President Eisenhower spoke as follows on a nationwide campaign program:

> I've got the best announcement that I think I can possibly make to America tonight. The progress made in the settlement of the Suez dispute this afternoon at the United Nations is most gratifying. . . .
>
> It looks like there's a very great problem that's behind us.

Stevenson had been careful in his campaigning—too careful for many Democrats—not to complicate Secretary Dulles' efforts to settle the Suez problem. At an overflowing press conference in Washington on September 17 he had explained his position:

> But this [the Middle East] is an area of vital concern to us and to our allies and I do not think that any comment or criticism by me at this crucial moment would serve a constructive purpose. No matter how the crisis arose, no matter whether it might have been avoided, we face a dangerous situation. And we all hope, regardless of domestic politics, that a peaceful solution can be found which preserves the unhampered use of the Canal and respects the rights of all concerned. We all hope, too, that our allies will find us resolute, reliable and fair-minded, as we have found them when matters of great importance to us were at stake.
>
> I do not want to add to the difficulties of the President

and the Secretary of State in this delicate situation. I shall reserve any further comment on this until a more appropriate time.

But when the President began to report "good news" from Suez, Stevenson concluded that his duty not to disturb the administration's efforts was transcended by an obligation to tell the people the truth. At Cincinnati, on October 19, he answered Eisenhower:

> We need to be called to labor, not lulled with rosy and misleading assurances that all is well. Leadership which fails in this is leadership to disaster.
>
> Yet a few nights ago the Republican candidate sought to make political capital out of a crisis that could engulf the world. Wars have begun over matters of far less moment than the Suez dispute—for the canal is a lifeline of the world.
>
> I have refrained until now from commenting on the Suez crisis. But the Republican candidate has introduced it, in a highly misleading way, into the campaign.
>
> A week ago he came before that so-called press conference on television arranged by advertising agents of the Republican campaign evidently more for adulation than for information. He announced that he had "good news" about Suez.

Angrily he denounced the President's attitude:

> But there is no "good news" about Suez. Why didn't the President tell us the truth: Why hasn't he told us frankly that what has happened in these past few months is that the Communist rulers of Soviet Russia have accomplished a Russian ambition that the Czars could never accomplish?

Russian power and influence have moved into the Middle East—the oil tank of Europe and Asia and the great bridge between East and West?

His next words hung with bitter sarcasm:

When the historians write of our era they may, I fear, find grim irony in the fact that when Russian power and influence were for the first time being firmly established in the Middle East, our government was loudly, proudly proclaiming our victorious conduct of the cold war and the President reported good news from Suez.

Six days later the Egyptian, Syrian, and Jordanian governments announced that they had placed their armies under a joint command. On October 29 Israel invaded Egypt. On October 30 Britain and France presented both Egypt and Israel with an ultimatum to withdraw from the canal area and lay down their arms within twelve hours. Israel agreed but Egypt refused. On October 31 Britain and France began air attacks on Egyptian military installations.

In the United Nations, meeting in emergency session, the United States joined with the Soviet Union to condemn the aggression of Israel, Britain, and France, and to demand a cease-fire. Thus the United States, under Eisenhower's contradictory policy of anti-communism on the one hand and impartiality on the other, found itself aligned with the Communist bloc against its own allies—and on an issue of war. On the night of October 31 Eisenhower addressed the nation on the emergency:

Tonight I report to you as your President.

We all realize that the full and free debate of a political campaign surrounds us. But the events and issues I wish to place before you this evening have no connection whatever with matters of partisanship. They are concerns of every American—his present and his future.

There could be no legitimate quarrel with the President's sentiments, nor any question that the crisis concerned everyone, regardless of party. But this opening was, once more, a masterful political stroke. Instead of going directly to the crisis, however, Eisenhower dealt first with the upheavals taking place in the Soviet satellites, Poland and Hungary. Just over half of his address was concerned with these matters. His theme was optimism. The struggles of the Poles and Hungarians were, he seemed to suggest, successes of American policy:

> We could not, of course, carry out this policy by resort to force. . . . But we did help to keep alive the hope of these people for freedom.
> We have rejoiced in all these historic events.
> Only yesterday, the Soviet Union issued an important statement of its relations with all the countries of Eastern Europe. This statement recognized the need for review of Soviet policies, and the amendment of these policies to meet the demands of the people for greater national independence and personal freedom.
> The Soviet Union declared its readiness to consider . . . withdrawal of its forces from Poland, Hungary and Rumania.

Only hours after these words were spoken the tanks of the Red Army rumbled into Budapest to slaughter the freedom-fighters in the streets; and the dying voice of the Hungarian freedom radio desperately implored America to give assistance in a struggle which, the Hungarians said, Americans had encouraged.

Eisenhower then turned to the darker side of his "report." He quickly reviewed developments in the Middle East through the seizure of the canal by Egypt, and American efforts to settle the canal problem:

> In the United Nations, only a short while ago, on the basis of agreed principles, it seemed that an acceptable accord was within our reach.
>
> But the direct relations of Egypt with both Israel and France kept worsening to a point at which first Israel, then France—and Great Britain also—determined that, in their judgment, there could be no protection of their vital interests without resort to force.
>
> Upon this decision events followed swiftly. . . .

Now he came to the breakdown of the Western Alliance:

> The United States was not consulted in any way about any phase of these actions. Nor were we informed of them in advance. . . .
>
> We believe these actions to have been taken in error, for we do not accept the use of force as a wise or a proper instrument for the settlement of international disputes.
>
> To say this, in this particular instance, is in no way to minimize our friendship with these nations, nor our determination to maintain these friendships.

This, too, was politically effective with its suggestion that though the American people and their President had been betrayed, yet the President was willing to forgive. He gave assurances that the United States would not become involved in the fighting. American policy, he emphasized, was to bring about peace, and American reliance would be upon the United Nations:

> We took our first measure in this action yesterday. We went to the United Nations [Security Council] with a request that the forces of Israel return to their own line and that hostilities in the area be brought to a close.
>
> This proposal was not adopted because it was vetoed by Great Britain and France.
>
> It is our hope and intent that this matter will be brought before the United Nations General Assembly. There, with no veto operating, the opinion of the world can be brought to bear in our quest for a just end to this tormenting problem.

He closed his address on a note of unexceptionable moral prescription:

> . . . As I review the march of world events in recent years I am ever more deeply convinced that the United Nations represents the soundest hope for peace in the world. . . .
>
> There can be no peace without law. And there can be no law if we work to invoke one code of international conduct for those who oppose us, and another for our friends.

Stevenson immediately requested, and received, equal free time the next evening to answer Eisenhower. He too "recited the record," briefly and indignantly:

> We not only failed to stop the introduction of Communist arms into the Middle East, but we refused to assist

Israel with arms too. We also refused to give Israel a guarantee of her integrity, although we had given such guarantees to others.

And in the meantime we dangled before Colonel Nasser the prospect of financial aid for building a great dam on the Nile.

In time, the bankruptcy of the Eisenhower administration's policy began to become evident even to Mr. Dulles. It became clear that Colonel Nasser was not a bulwark of stability, but a threat to peace in the Middle East. Thereupon President Eisenhower abruptly and publicly withdrew the aid he had led Colonel Nasser to expect.

As anyone could have foreseen, Colonel Nasser promptly retaliated by seizing the Suez Canal.

Driven by our policy into isolation and desperation, Israel evidently became convinced that the only hope remaining was to attack Egypt before Egypt attacked her. So she took her tragic decision.

In a biting paragraph Stevenson summed up the situation, four days before the American people were once again to choose between him and Eisenhower:

Here we stand today. We have alienated our chief European allies. We have alienated Israel. We have alienated Egypt and the Arab countries. And in the U.N. our main associate in Middle Eastern matters now appears to be Communist Russia—in the very week when the Red Army has been shooting down the brave people of Hungary and Poland. We have lost every point in the game. I doubt if ever before in our diplomatic history has any policy been such an abysmal, such a complete and such a catastrophic failure.

Though perhaps not "anyone" could have foreseen Nasser's seizure of the Suez Canal, this was in the main a just indictment. Stevenson went on to recall his own proposals for Middle Eastern policy made a year before and repeated often thereafter. His position as a candidate precluded his making such specific suggestions as might inhibit the efforts of the administration at the U.N., but the repetition here of his consistent policy suggested clearly enough the course he thought should be followed. At the end he sought to take as positive a stand as circumstances would permit:

> I would not condone the use of force, even by our friends and allies. But I say that we now have an opportunity to use our great potential moral authority, our own statesmanship, the weight of our economic power, to bring about solutions to the whole range of complex problems confronting the free world in the Middle East.
>
> The time has come to wipe the slate clean and begin anew. We must, for a change, be honest with ourselves and honest with the rest of the world. The search for peace demands the best that is in us. The time is now. We can no longer escape the challenge of history.

Whether a majority of Americans, disagreeing with Stevenson, thought that the "challenge of history" could still be "escaped," or whether they thought it best to entrust their affairs to a military man in a time of war and crisis, or whether, indeed, they simply kept their preference of Eisenhower to Stevenson despite the record being written almost hourly—that majority, in even larger numbers than before, returned Eisenhower to the White House four days later.

But in the United Nations, whatever the American voters may have thought, the challenge of history could not be escaped. On November 3 Lester Pearson, Canadian Minister for External Affairs, moved that the United Nations establish immediately a police force to take over patrol of the Middle Eastern borders as soon as a cease-fire could be achieved. On November 6, the day after the American election, Britain, France, and Israel agreed to a cease-fire with Egypt. Within hours the United Nations' forces took over. At the moment of his defeat, Stevenson's policy became world policy, not because it was his but because it was necessary. The tragedy was that the rejection of his policy had made the outbreak of the war possible, with all its disastrous moral and political consequences. As he returned again to private life the irony of his career in national politics was complete. A majority of Americans, complacent or trusting, preferred Eisenhower as the symbol of their wishes for noninvolvement; yet the circumstances required the policies of involvement and intervention which Stevenson had offered in his articulation of realities.

## CHAPTER V

# Stevenson in Washington

STEVENSON had returned to private life and to his law practice when the first Russian sputnik was launched into space in September, 1957. This historic demonstration of Soviet scientific advancement and military potential brought a quick end to the complacent air with which the Eisenhower administration had appeared to be viewing the world scene. British Prime Minister Macmillan flew to Washington for conferences with President Eisenhower, who took to television to reassure the nation and to call for a greater educational effort in the field of science. The President and Secretary Dulles, in hasty conference with British and other allied leaders, determined to convert the fall NATO meeting, set for December in Paris, into a conference of allied heads of state. That meeting immediately assumed a character of importance and urgency unparalleled since the Korean War. United States proposals would have momentous consequences.

Late in October, 1957, after the successful launching of the second and much larger sputnik, Secretary Dulles persuaded President Eisenhower that some dramatic step should be taken to demonstrate American unity in the face of the mounting Soviet challenge. It was decided to invite Adlai Stevenson, as titular leader of the opposition party and a

veteran of foreign affairs, to participate in formulating the
American program for NATO and to take part in the NATO
conference at Paris.

Both the approach to Stevenson and his initial response
were hesitant and somewhat ambiguous. The force of Secretary Dulles' invitation was that Stevenson should take the
lead in preparing the American position, make full use
of the State Department staff, and then present his formula-
tions to the Secretary and the President to use as they saw fit.
As a staff official under the Secretary of State he would have
no authority to match his responsibility. It was clear that the
administration wanted Stevenson's help, both because of his
position at home and because of the great esteem in which
he was held abroad. But it was also clear, and understandable,
that the administration was not anxious to have the defeated
Democratic leader, rather than the President, receive credit
for a successful meeting of NATO. As for Stevenson, he
was willing to help and felt a strong sense of duty to do what
he could. But as Democratic leader he could not commit his
party in advance to a program it might disapprove, nor could
he commit himself to accepting responsibility for policies he
would have no authority to put forward or with which he
might not agree. Some of his colleagues among the Democratic leadership felt strongly that he should not, in any case,
accept a post below that of the Secretary of State. The
problem was resolved after considerable consultation on
both sides. Stevenson sought the advice of former President
Truman, Senate Majority Leader Lyndon Johnson, Speaker
Sam Rayburn, colleagues on the Democratic Advisory Council, and several long-time associates. The *modus vivendi* to
which he agreed was that he would go to Washington as a
consultant only, would review the American position papers
as they were developed, and give his advice throughout.
Where he found himself in agreement with administration

proposals he would give his support, but he would remain free to criticize. Whether he would go to the Paris conference was left open. A letter from Stevenson to President Eisenhower, dated November 17, contains a good statement of the terms:

> Perhaps I should take this opportunity to say what I am sure we all understand—that while I must be free to seek advice, in my informal consultative capacity, from persons outside the Department, including leaders of my party, and also to express my views, even where they may differ from the Administration, I shall strive to promote national unity in furtherance of the great tasks before us.*

At this stage Stevenson was already at work. In response to Secretary Dulles' initial request for assistance he prepared a preliminary memorandum for limited circulation in the State Department. His point of departure was the Declaration of Common Purpose issued by President Eisenhower and Prime Minister Macmillan after their meeting on October 25. "The emphasis of this communiqué," he wrote, "was almost exclusively military." He was willing to accept the military as having priority but, he said, it was "by no means the whole answer." He concerned himself with positive suggestions to buttress the objectives announced by Eisenhower and Macmillan. These words sounded the characteristic note both of this first memorandum and of Stevenson's whole NATO mission:

> I am not sure that NATO is a suitable instrument for collective economic development programs in the retarded

* I am grateful to Governor Stevenson for permission to quote from this letter and from documents prepared by him during his NATO mission.

countries—a responsibility and mechanism of defense that is of equivalent or greater importance than military strength.

But the NATO Council should be able to devise machinery for prompt consultation and action in cases involving member states, like the Iceland fish and Lebanon apples.

Again:

The main threat is *not military aggression,* but subversion by propaganda, economic bribery and political penetration. Have we any common plans to counter such ambiguous aggressions?

And again:

If the Atlantic Community had multilateral economic and trade development plans it would mean a lot more to many people than its purely military anti-communism does now.

Is an experimental plan to stabilize some raw material prices which is so important to the underdeveloped areas, beyond the capacity of a league of the principal industrial states like NATO?

In the same preliminary memorandum Stevenson re-emphasized his belief in the necessity for active steps toward disarmament:

We must be ready for new Russian proposals which may be extremely effective as propaganda. I have often said that suspension of nuclear testing with suitable monitoring posts to safeguard against violation should not be made conditional on cessation of production of nuclear materials. It would be an important first step; it would break the deadlock and halt or slow down the dangerous race which at best leads only to stalemate and a balance of terror.

For several weeks, at the end of November and early December, Stevenson worked at the State Department in Washington. His presence there was greeted with almost universal approval by the press both at home and abroad among the Western allies. It was precisely the symbol of national unity the administration sought. If the unity thus achieved was more apparent than real, it was certainly better than bitter division, and NATO was assuredly strengthened by it.

But below the surface in Washington the administration had little enthusiasm for a drastic reappraisal of its policy and program. What was wanted of him, so it seemed to Stevenson, was simply approval of what had been going on all the time. "I am troubled," he wrote in a memorandum of November 29 to Dulles, "by the lack of a sense of urgency. I came to Washington to work first in 1933, again early in 1941; both times the atmosphere was different. I wish it was now." He went on:

The response to Sputnik, etc. doesn't seem to meet the measure of the emergency. It seems to be compounded largely of more missiles and more reassurances. But it is obvious to the informed that a much greater effort is required all along the line, and that firm and far-reaching decisions in principle should be made promptly within the Executive Branch. If it is made clear now that Congress will be asked for a comprehensive program to reverse current trends, I believe public and Congressional support would be better secured.

This basic decision is related to NATO. I doubt very

much if it will be possible to communicate much sense of
urgency and determination to our allies in Paris if we have
not made the measure of the emergency clear at home.

Stevenson urged the closest political relations with NATO
and firm assurances to the NATO partners that consultation
with them was fundamental to American policy. The following passage not only shows clearly the line Stevenson proposed, but the attitude of the administration he found it
necessary to counter:

> You have stressed the dilemma of choice between close
> collaboration with our Atlantic allies and winning support
> among Asian and African states. But I do not believe that
> we have interests in the Middle East and in Asia which
> differ materially any longer from those of our NATO part-
> ners. I doubt whether our interest in the survival of a
> democratic India is any greater or any less than Europe's.
> Collapse of the Western position in the Pacific would be
> as fatal to Europe as to us. Who can say whether we or
> our NATO partners have been more seriously damaged by
> the deteriorating situation in the Middle East? Our differ-
> ences with our allies have not been so much differences
> of interest as differences in judgment as to the wisdom of
> actions.

Specifically, Stevenson advised that NATO should be employed for the solution of "enfeebling internecine conflicts,
like Algeria, Cyprus and Tunisia." These, he said, were
problems not only for the British and the French but for
all the partners. "Similarly," he continued, "our problems
with South Korea, Taiwan, Jordan, etc. are problems in which
our NATO allies suffer with us." It is plain from the context
of these passages that Stevenson found the administration

more than reluctant to discuss such matters, though they seemed to him to suggest not only points of utmost importance for the alliance but also to reveal a chief source of weakness in American policy. "I know your concern about flexibility for prompt action," Stevenson wrote, "but I do not think it incompatible with a greater anxiety to harmonize political action than the papers for the meeting suggest." The administration apparently continued to think of "flexibility" as incompatible with such close political relations. At any rate, despite the lessons of Suez, no action in this direction was taken at the conference.

Next Stevenson turned briefly to the military aspect of NATO's problem, to emphasize his concern lest the alliance be stripped of capability to fight any but a nuclear war. While he clearly did not conceive his role as involving advice on military questions, he felt it necessary to express a "caveat," as he called it, with regard to the apparent "intention to use tactical nuclear weapons as the automatic answer to Soviet aggression." He offered these observations:

> The Soviet Union is more likely to create an ambiguous than an unequivocal aggression. What we may well be confronted with on the Eastern front is a challenge to Berlin or to Yugoslavia or to Warsaw—a challenge resulting not so much from Soviet tanks or guns as from a subversive local Communist movement. I am not at all sure that such an "ambiguous aggression" can be adequately met by the forces now available to the Western Alliance.
>
> At the same time, can any aggression be met by the use of tactical nuclear weapons, in a crowded industrial area such as Western Europe, without precipitating general war?

At the 1957 NATO meeting little was said about the problem Stevenson was raising, and no action was taken. But in the spring of 1959, when the Berlin crisis flared, both President Eisenhower and General Lauris Norstad, NATO commander, conceded that NATO could not stand off the Soviets in Europe by means of conventional arms, that if war came over Berlin it would be nuclear war.

Stevenson, in his memorandum of November 29, gave attention chiefly to the question of economic development, enlarging on the view he had stated at the outset. He began by stating flatly that he found the "suggested position paper on economic assistance inadequate." Since, as he put it, "the hottest war is the cold war," it followed that nonmilitary programs would be of at least equal importance with military:

> The purpose of the meeting in Paris is to strengthen not only NATO but the free world as a whole. We should make it clear that in the American view the military defense of Europe and the winning of the economic battle for the improvement of the conditions of life of the uncommitted peoples are not alternative imperatives for the NATO countries—they are both necessary. Moreover, the tasks of military policy are negative and an insufficient expression of the common aspirations of our peoples and of peoples throughout the Free World.

The depth of Stevenson's feeling on this crucial matter is suggested by the fact that he departed from his role as critic to make extensive proposals for a free world economic policy and an American program. He listed his recommendations as follows:

(1) Announce that we are asking Congress for a substantial enlargement of our foreign development funds.

(2) Call upon the other NATO countries to make increased resources available for underdeveloped countries.

(3) Propose a special meeting of the OEEC [Organization for European Economic Cooperation] to be held in January to

(a) determine a scale of effort for foreign economic development; and

(b) improve coordination among the participating countries.

The advantages of such a program would be great, Stevenson argued, since it would "moderate the impression that the NATO meeting was concerned almost exclusively with military matters," and since it would "reassure underdeveloped countries that the West was mindful of their first concern (which is *not* the Soviet threat)." While he wished NATO to launch a co-ordinated program of economic assistance, he was careful not to limit the participants to OEEC. Specifically he suggested that such countries as Switzerland, Canada, Australia, New Zealand, Argentina and Cuba might be invited. After preliminary plans had been laid out at a meeting of OEEC, he would call a general meeting of all interested governments. As for the American contribution, Stevenson thought that perhaps five hundred million dollars additional should be made available for development loans, of which 40 to 50 per cent might be earmarked for India.

Turning to the related problem of trade policy, Stevenson offered suggestions directed toward breaking down economic walls as rapidly as possible and committing the United States both to support of the European common market and to extension for five years of its own reciprocal trade program.

In view of the special importance he attached to the political and economic functions of NATO as distinct from or parallel to the military, Stevenson now thought it useful to redefine the meaning of NATO itself. "The exclusively military emphasis in NATO has not enhanced respect in the cold war areas," he wrote. "It would be well to vigorously call attention to its wider meaning and potential significance." Since none of the papers being prepared for the conference seemed to do this, Stevenson wrote out his own views at some length. Since this passage is the heart of his November 29 memorandum and reveals most clearly the concern which led him to go to Washington, it is worth extended quotation:

NATO is more than a temporary military alliance. Although it was born of the urgent necessity to defend our free institutions and cultural traditions, the roots of our great community of interest are much deeper than common defense. And the community of interest will outlive the threat which brought NATO into being.

We have reached the stage where the historical developments which unify our nations—our recognition of the dignity of the individual, the responsibility of government to the people, the universality of certain basic human rights, and the love of God embodied in our religions—must begin to give us the strength to forge an even greater unity of purpose. The atom and the supersonic missile have made us realize our proximity and our vulnerability. We must now seize the opportunity to agree on goals and find common paths to reach them. . . .

NATO commits North America permanently to sharing in the solution of the problems of the Atlantic Community and the reconciliation of differences between its members. And beyond its frontiers, NATO must serve to extend to others the advantages of the modern industrial society which we enjoy. . . .

That we have not realized the greater values for mankind in this closer association of the Western community is due to the relentless Soviet military threat. But in solving the immediate problems which face us in the military and the political arena, we must not lose sight of the fundamentals which give meaning to our great Community. For it is these fundamentals which make worth-while the many sacrifices which we have had to make, and will have to make, to defend our civilization and all that it stands for.

Holding such views about NATO, Stevenson could not approve the draft of the President's speech for the conference, with its heavy emphasis on military capability. It lacked "heart," Stevenson thought. When invited to work on the speech, Stevenson tried to put some "heart" into it. He also argued for putting into it some sense of the urgent need for disarmament. Recalling his long-standing belief that the West should favor suspension of nuclear tests without qualifying riders, he again expressed the hope that the United States, through NATO, would take the initiative away from the Soviet Union in this vital sector of policy. "I would carefully refrain from any appearance of obstinacy about our present proposals and welcome a resumption of negotiations and new initiatives," he wrote. "Indeed, I would, as you know, offer some ourselves!"

Stevenson's final suggestion was that the closing communiqué of the conference ought to conclude "with a brief and ringing statement summing up NATO policies in simple language—'Arms only for defense. Aid for the needy, underdeveloped nations. Cooperation with all states, including the Communist nations, to promote world peace and progress.'"

Stevenson's repeated emphasis on economic development finally elicited a request from Secretary Dulles that he explain his views further. He did so in a detailed memorandum dated December 5. In this paper he speaks of an official position

paper which is an "expression of 'interest' in an enlargement of the resources available to the 'less developed areas.' "
But, he says, "the *action* portions which I suggested have
been deleted." They remained deleted despite Stevenson's
pleas for a positive approach to supplement and parallel the
military.

In the December 5 memorandum Stevenson pointed out
that "it is manifest from the figures and the facts on Egypt,
Syria, Afghanistan, and elsewhere, not to mention the rising
Chinese pressures in Asia, that the political and economic
penetration of the Communist bloc must be arrested." The
situation called, he thought, "as imperatively for a combined
Western effort as the Soviet bloc's weapon capability." He
put his proposals succinctly in this passage:

> The sensible way to proceed is to establish existing
> levels of effort as a floor; and commit all participating
> industrialized nations to an increase. There will undoubt-
> edly be a certain amount of fun and games in switching
> funds from "existing" to "additional" categories; but if a
> serious commitment is made in principle, the bookkeeping
> can probably be handled.
>
> I fear the tendency for Western Europe to channel all
> capital exports into residual colonial areas—and to induce
> Germany to back this effort—is not altogether wholesome.
> The present proposal represents an occasion to begin the
> gradual move of more Western capital to, say, the Col-
> ombo Plan area, and more American capital into Africa.
>
> With respect to Germany and—perhaps—with respect
> to other Western European countries, the United States
> ought to be prepared to bargain hard for an increase in
> capital exports under the proposed arrangement. Because
> others, like ourselves, are suspicious of international funds,

it will have to be made clear that what is contemplated is the coordination of sovereign national efforts.

The mechanics of the program would involve use of the facilities of OEEC, enlarged to include several non-NATO nations having capital to export. "Too often," Stevenson wrote, "I suspect the absence of suitable machinery has been a perfect excuse for the people who, for financial or other reasons, want to do nothing." OEEC had, he pointed out, the advantages of being established, with a secretariat, being identified with economic rather than military functions, and being capable of expanding effort. Beyond the initial pooling efforts of OEEC Stevenson envisioned the use of that agency for Western co-operation with and support of a joint Middle East development agency, "as a logical extension of economic cooperation into the Mediterranean area and a demonstration of our willingness to commit technical and financial resources to sound development without political strings." In turn this agency, he thought, might develop the Jordan River valley, assist in resettlement of refugees, and help to channel oil revenues into "regional development projects."

Stevenson concluded this paper, his final formal memorandum for Dulles, with these words:

I hope very much that at NATO the U.S. can express something more than "interest" in the less-developed, uncommitted areas where the cold war is hot. I can think of nothing more effective than a serious, thoughtful proposal to mobilize the resources and skills of the democratic "haves" for the development of the "have nots."

The ambiguity of Stevenson's position in Washington continued until the end. The President's invitation to him to go to Paris was, to say the least, less than direct. Stevenson decided that he had done all that he could, that to go to Paris could serve no useful purpose. The actual positions of the United States at Paris reveal clearly enough why the invitation to Stevenson was halfhearted and why he could not go. The President and the Secretary of State had decided to take no new positions but rather to concentrate all their energies upon assertion of American military might as more than capable of offsetting the Russian, and upon persuading NATO allies to accept installation of intermediate-range ballistic missiles.

When their speeches were delivered at Paris, all that remained of Stevenson's curious adventure among his long-time political opponents was a kind of ghostly reminder of his hope that the President's principal speech might commence with something positive and peaceful rather than military:

> This is a time for greatness.
>
> We pray for greatness in courage of will to explore every path of common enterprise that may advance the cause of justice and freedom.
>
> We pray for greatness in sympathy and comradeship that we may labor together to end the mutual differences that hamper our forward march within a mutual destiny.
>
> We pray for greatness in the spirit of self-sacrifice, so that we may forsake lesser objectives and interests to devote ourselves wholly to the well-being of all of us.
>
> We pray for greatness of wisdom and faith that will create in all of us the resolve that whatever measures we take, will be measures for peace!
>
> By peace, I do not mean the barren concept of a world where open war for a time is put off because the com-

petitive war machines, which humans build, tend mutually
to neutralize the terrors they create.

Nor by peace do I mean an uneasy absence of strife
bought at the price of cowardly surrender of principle.
We cannot have peace and ignore righteous aspirations
and noble heritages.

The peace we do seek is an expanding state of justice
and understanding. It is a peace within which men and
women can freely exercise their inalienable rights to life,
liberty and the pursuit of happiness.

In it mankind can produce freely, trade freely, travel
freely, think freely, pray freely.

The peace we seek is a creative and dynamic state of
flourishing institutions, of prosperous economies, of deeper
spiritual insight for all nations and all men.

If the voice was Eisenhower's, the words nevertheless con-
veyed in some measure the spirit Stevenson had wished to
invoke for the conference. But the little success was lost in
the big failure. Neither at Paris nor afterward was American
effort devoted to augmenting NATO for positive, nonmilitary
purposes. In a press conference a few days before the Paris
meeting Secretary Dulles was asked what he thought of the
proposal, advanced by Italian Foreign Minister Pella, that
a fund administered through OEEC might be used to aid in
the development of the Middle East. Fresh from his rejection
of the same idea as proposed by Stevenson, Dulles answered:
"I was about to say that in principle I thought it was good.
It is a useful suggestion and the concepts that underlie it
are quite acceptable to the United States." At Paris President
Eisenhower, exactly as Stevenson had feared and foretold,
expressed an "interest" in economic development. "We have,"
said the President, "been parties to the grant of political

liberty to hundreds of millions of people. But that bestowal could be a barren gift, and indeed one which could recoil against us, unless ways are found to help less developed countries to achieve an increasing welfare." But this was all he said, and nothing was done or proposed for more than two years. American economic assistance continued to be given in largest measure where it supported military positions against communism, and American policy continued to temporize with the problem of the uncommitted world. NATO had no development policy at all. It became increasingly clear by 1960 that liberty would indeed be a "barren gift" in many countries if the free world failed to make possible for them the economic development they were being offered by the Soviet Union. When in December, 1960, Under Secretary Douglas Dillon, on behalf of the United States, signed the articles setting up the Organization for Economic Co-operation and Development, another Stevenson proposal was belatedly adopted by its detractors.

## CHAPTER VI

# Defense and Disarmament:

# 1956 and After

I

BEFORE the eruption of war in the Middle East, two issues of the 1956 Presidential campaign seemed to attract more popular attention and concern than any others. These were Stevenson's proposals for new policies of military recruitment, looking to an end of the draft, and his proposal to stop the testing of hydrogen bombs. Both issues were sharply debated by the candidates and in the press but, like all other issues, they were largely forgotten in the final, hectic days when the eyes of America and the world were focused on Suez. Both nevertheless remained to be settled, regardless of the outcome of the election.

Stevenson first raised the question of the draft in his address to the American Legion Convention at Los Angeles on September 5:

Many military thinkers believe that the armies of the future, a future now upon us, will employ mobile, technically trained and highly professional units, equipped with tactical atomic weapons. Already it has become apparent that our most urgent need is to encourage trained men

189

to re-enlist rather than to multiply the number of partly trained men as we are currently doing.

For these reasons, he suggested, a better means of military recruitment might well replace the draft in the "foreseeable future." Immediately the news was full of the draft question. It looked for a moment, at least, as though Stevenson had hit upon a popular note. To some it even seemed a cheap one. Like all politicians Stevenson suffered from quotation. The idea that the draft might be ended was seized upon; the reasons and alternatives ignored. When asked if he saw any chance of ending the draft in the foreseeable future, the President quickly responded, "No, I don't." These words fixed the Republican campaign position. A few days later, September 19, Eisenhower expanded his view:

> It [ending the draft] would weaken our armed forces. It would propagate neutralist sentiment everywhere. And it would shock our allies who are calling upon their people to shoulder arms in our common cause.

Such a course, said the President, would be to "face the future by walking into the past backward."

Stevenson, nettled by the "authoritative" manner in which his suggestion was dismissed, and ignoring the advice of some of his friends that he could not hope to win an argument with Eisenhower on a military question, returned to the matter on September 29 at Minneapolis:

> . . . I have said before and I'll say it again that I, for one, am not content to accept the idea that there can be no end to the draft, to compulsory military service.
>
> Let me make it perfectly clear that, as long as danger confronts us, I believe we should have stronger, not weaker, defenses. . . .

But my point is that the draft does not necessarily mean a strong defense. Conditions change, and no conditions have changed more in our time than the conditions of warfare. Nothing is more hazardous in military policy than rigid adherence to obsolete ideas. France crouched behind the Maginot Line, which was designed for an earlier war, and German Panzers ran around the end. The Maginot Line gave France a false—and fatal—sense of security. We must not let Selective Service become our Maginot Line.

Next he tried to place the whole discussion on a more positive footing, and repeated his suggestion in a comprehensive manner:

What I am suggesting is that we ought to take a fresh and openminded look at the weapons revolution in connection with the whole problem of recruiting and training military manpower. We may very well find that in the not-far-distant future we not only can but should abolish the draft in order to have a stronger defense and at lower cost. Defense is now so complex, its demand for highly skilled and specialized manpower so great, that the old-fashioned conscript army, in which many men serve short terms of duty, is becoming less and less suited to the skilled needs of modern arms. And it is becoming more and more expensive.

Thus did the civilian candidate presume to read the military President a lecture on military manpower. On October 6 Eisenhower took up the challenge briefly. Stevenson's draft talk, he said, was "loose talk." To compare the draft with the Maginot Line showed "an ignorance of military needs or a disregard for national security." The draft, the President concluded, had proved "indispensable" in holding the armed forces at maximum strength and in spurring voluntary enlistments. For most editors, at least, this statement seemed to be enough.

But rumors spread that Eisenhower had himself planned to announce, toward the end of the campaign, that the draft was to be ended, and that Stevenson thus had "scooped" him. On October 12, when a reporter suggested this possibility, the President responded with irritation. He cited his record on national security, repeated his insistence on the draft, and announced that he had said his "last word" on the subject.

However, the last word had not been spoken, either by Stevenson or Eisenhower. Against urgent political advice, Stevenson made the draft question the theme of his speech at Youngstown, Ohio, on October 18. Refusing to be put off either by the President's pronouncements or the chorus of editorial opposition, he asserted: "It is a serious subject and I must speak seriously." First he reviewed his earlier statements on military manpower procurement and the draft system, and his suggestion for a reappraisal of the whole situation:

> This suggestion has been taken by some—and deliberately misconstrued by others—as a proposal for weakening our armed forces. It is exactly the opposite. It is a proposal for strengthening our armed forces.
>
> The point is simply that we already need and will need more and more a type of military personnel—experienced and professional—which our present draft system does not give us. The draft means a tremendous turnover in our military personnel, and a resultant high proportion of inexperienced personnel. There is ample evidence that this inexperienced personnel is not meeting today's needs.

To support his position, he cited the views of the Army's Deputy Chief of Staff for Personnel, and the Air Force Chief of Staff. But this was a political rally, not a war college seminar, and Stevenson now strove for as much political appeal as he could make:

> Every young man who has served in our armed forces knows the incredible waste of our present system of forced but short-term service. He knows the money that could be saved, the new efficiency that could result from a volunteer system which calls on young men not to endure two years of service because they have to, but to choose it—and for a longer period—because it offers advantages that seem to them appealing.

What advantages could be offered? Stevenson now combined his political appeal with positive proposals:

> There seems to me every reason for searching out ways of making military service attractive enough that sufficient numbers of young men will choose such service voluntarily and will then remain in the services for longer periods.
>
> By cutting down the turnover we can reduce the present enormous cost of training replacement after replacement. The money that is saved by this reduction in training costs can be used to pay our soldiers, sailors and airmen better salaries, to provide them with improved working conditions and perhaps to offer special bonus inducements for longer service. In this way we can develop a more effective defense, with higher morale, and I believe no higher cost.
>
> Where there are needs for particularly highly trained men, as for example in radar, electronics and other spe-

cialties, I think we should consider offering university scholarships which will provide specialized training, in conjunction with a liberal education, to applicants, otherwise qualified, who will agree to spend a specific period in the armed services.

At the end Stevenson endeavored to tie these proposals together with their political value by appealing from the General in the White House to the people:

. . . I think I speak for every person in America—that we will count it a better day when we find that these military needs can best be met by a system which does not mean the disruption of the lives of an entire generation of young men; which lets them plan their education, and get started more quickly along life's ordained course.

This, I submit, is a matter that should be seriously considered by the American people. The Republican candidates insist that it should not even be discussed, that this isn't the people's business, and that with a military man in the White House things like this can best be left up to him.

Well, I say just this: What is involved here is the security, perhaps the life or death of this nation. What is involved here is the use that should be made of two years of our sons' lives. What is involved here is whether there should be new ways of more effectively meeting new problems. And I say that these are decisions that must be made not by one man—not by one general—not even by one man as President—but by the American people.

Vice President Nixon responded by comparing Stevenson to a bush league ball player trying to break into the majors. And Eisenhower found that he had another word to say after all. Stevenson's views, he said, were "incredible folly," and his proposals led "down the road of surrender." While the press headlined the strong language of Eisenhower, Stevenson persisted and repeated his views the next day at Cincinnati.

But Stevenson's political advisers, and the editors of the nation's newspapers, were right on one score. The election returns showed clearly enough that the majority of the people preferred the judgment of the President in military as in other matters. The civilian candidate could not hope to win his argument on a military matter at the ballot box.

## II

To win an election at the polls is one thing; to give effective leadership in the business of government may be another. The problem of military manpower, like other great issues of the Eisenhower era, could not be settled by ignoring it, by abrupt dismissal of new ideas, or by appeal to the prestige of a hero-President. Stevenson, the politician and candidate, had laid out the essentials of an urgent problem and pointed the way to resolving it. When the election was over and the heat of politics had cooled, the Department of Defense began quietly to revise its manpower procurement system.

In May, 1957, a special government committee appointed by the Secretary of Defense—the Defense Advisory Committee on Professional and Technical Compensation—made its report to the Secretary of Defense. The committee, headed by Ralph J. Cordiner, president of the General Electric Company and a strong supporter of President Eisenhower, and including a distinguished list of civilian and military members expert in personnel problems, entitled its report *A Modern Concept of Manpower Management and Compensation*. From beginning to end it read like a documented elaboration of Stevenson's proposals. Here are some examples of the committee's findings and advice to the Secretary of Defense:

> Technological change means a change of weapons in the combat units, change in the techniques required in weapons maintenance and use, and change in the level of skill and judgment in the user. The day has passed when a large portion of the military workforce performed relatively unskilled tasks and a major measure of their competence was based upon discipline and physical fitness only. Today, a large portion of the defense team must possess not only the discipline and physical and mental stamina formerly required but also a trained, experienced and disciplined skill in the use of complex equipment.

> Development of this level of skill requires adequate quality input to training, an intensive program of instruction and supervised on-the-job experience before the full effectiveness of the individual and equipment in his charge can be realized.

> Man is still the primary element of defense. It is he who causes the newer, more complex, more potent weapon to respond promptly and deliver its full potential with accuracy. Without the control of the skilled individual the weapon is only an inert, complicated and expensive device.

> The time and effort required to impart the training and experience necessary to control with maximum effectiveness the more potent, more complex weapons of today have markedly increased. This is true notwithstanding the generally rising level of education of our national manpower. Obtaining a proper return for this increased training effort requires that the services of the trained and experienced individual be retained for a reasonable period of productive service.

> Such retention is not being realized today to an accept-

able, economic degree. It is least realized in the skills requiring the most lengthy and costly training. Today there is a tremendous outflow of effort to train a stream of transient personnel to a journeyman level of competence without a reasonable realization of skilled service in return. This is the heart of the enlisted retention problem.

The reasons for this failure are numerous but basically relate themselves to a comparison of the total emoluments of voluntary military service with those available to the same quality of manpower in the civilian economy. *The quality and degree of retention of skilled manpower required by the Services cannot be secured by compulsion in a democratic society at peace.* Service of the caliber required cannot realistically be expected to flow primarily from patriotic motivation, felt by the small segment of society involved. An acceptable degree of retention of quality manpower in peacetime military service can be secured in a free society only by according those concerned a reasonable measure of the prestige and benefits they could otherwise achieve in civil pursuits in the mainstream of the economy.*

Here was Stevenson's central theme. A thorough, technical survey of military manpower problems, such as he had himself suggested, had been made, and its findings were those he had anticipated.

In other portions of the report the committee made detailed suggestions for pay scales according to skills, for better housing, better insurance programs, and for university scholarships—all suggestions Stevenson had advanced during the 1956 campaign. In sum, the advice of the Eisenhower administration's committee of experts was to follow the course

* Pp. 43–44. Italics mine.

Eisenhower himself had called "incredible folly." It is interesting, if not very important, to notice that the committee carefully avoided using the word "draft." Perhaps they concluded that discretion was indeed the better part of political valor—and "compulsion" would do just as well! But when the committee report was made public the President's irritable treatment of the whole draft question proved to have established obstacles of real consequence. The Department of Defense found it inexpedient to make a sustained effort to obtain congressional support for its new program, and members of Congress were not eager to be of assistance. Nor could the Cordiner recommendations be easily adjusted to an "economy" budget for defense. Instead of outright adoption and forthright action, the new program has been put into operation only gradually and without publicity. And the draft was permitted to diminish with disuse. It was important to save the victorious nonpartisan President from embarrassment over his "political" handling of a nonpartisan issue—while the proposals of his defeated and partisan opponent became national policy.

By the end of 1957, with sputniks one and two speeding around the earth, the President's embarrassment seemed less important, even to his leading supporters. *Life* was disturbed by the slow progress in putting the military on a thoroughly modern basis of competence. "We Need Pros in the Services," headlined its editorial of November 25:

> Thanks largely to General Eisenhower's long advocacy, universal peacetime service was finally accepted and enacted in 1951. But it was obsolete as soon as we got it. The cadres trained for atomic weapons and tactics will be the indispensable part of any operational army, however huge. But under the present system, the best trained officers and technicians are lost to the service just as their

training is beginning to pay off. The Cordiner report recommended pay scales that would check this costly turnover. Since it would add $565 million to the military budget during the first two years, the Administration refused to put it in the 1957 budget. But the savings in subsequent years would soon be reckoned in billions. If this plan were now in effect, these savings would be a lot closer. So would a more efficient army, navy and air force.

*Life,* only a year before, had belittled Stevenson in comparison with Eisenhower; but neither *Life* nor the nation could continue to belittle Stevenson's ideas when realities made them imperative. When in 1959, on the heels of new war scares, the Congress extended the draft, it was clear to everyone that the only real purpose it served was to provide cheap common labor for the Army. The Navy, Marine Corps, and Air Force had not used selective service for years.

### III

Aside from the war in Suez, Stevenson's proposal to bring an end to the testing of hydrogen superbombs was the most spectacular issue of the 1956 campaign. After the firing along the canal had ceased, it remained perhaps the most pressing issue confronting the whole world. In retrospect it is odd that the H-bomb question, first raised by Stevenson in April, did not capture public attention until long afterwards. In September and October candidates spoke thousands of words about it and newspapers carried thousands of column inches. Some chose to think that Stevenson wished to frighten the voters into supporting him; others thought he was searching for an issue to counter the image of Eisenhower as peacemaker; Eisenhower himself called Stevenson's proposal a "theatrical national gesture." But the fact was that the effort to stop the testing of the superbombs was a central element in Stevenson's policy from the outset. He first dealt with it directly not at a political rally but in his address on foreign policy to the American Society of Newspaper Editors in Washington, April 21, 1956.

In that address he said that he recognized an "obligation to measure criticism by affirmative suggestion." The H-bomb proposal was the second of three he offered:

> . . . I believe we should give prompt and earnest consideration to stopping further tests of the hydrogen bomb, as Commissioner Murray of the Atomic Energy Commission recently proposed. As a layman I hope I can question the sense in multiplying and enlarging weapons of a destructive power already almost incomprehensible. I would call upon other nations, the Soviet Union, to follow our lead, and if they don't and persist in further tests we will know about it and we can reconsider our policy.
>
> I deeply believe that if we are to make progress toward effective reduction and control of armaments, it will probably come a step at a time. And this is a step which, it seems to me, we might now take, a step which would reflect our determination never to plunge the world into nuclear holocaust, a step which would reaffirm our purpose to act with humility and a decent concern for world opinion.

In view of the outrageous attempt of Soviet Premier Bulganin five months later to interfere in the American election, Stevenson's next paragraph takes on a note of special significance. He placed it in parentheses:

> (After writing this last week down south, I read last night in Philadelphia that the Soviet Union has protested a sched-

uled H-bomb test. After some reflection I concluded that I would not be intimidated by the Communists and would not alter what I had written. For this suggestion is right or wrong and should be so considered regardless of the Soviet.)

After President Eisenhower had addressed the same convention in the evening of the same day, the editors voted informally two to one that Stevenson was right in asserting that the United States was "losing the cold war," and Eisenhower wrong. But in their editorials the editors said next to nothing about the H-bomb proposal. A few days later, at his press conference, the President was asked specifically to comment on the proposal. He regretted that the reporter had quoted Stevenson, since "I don't comment on somebody else's opinion." Nevertheless he answered the question at length:

> But I do want to point this out. It is a little bit of a paradox to urge that we work just as hard as we know how on the guided missile and that we stop all research on the hydrogen bomb, because one without the other is rather useless. So we go ahead with the hydrogen bomb, not to make a bigger bomb, not to cause more destruction, but to find out ways and means to which you can limit it, make it useful in defensive purposes of shooting against a fleet of airplanes that are coming over, to reduce fallout, to make it more of a military weapon and less one just of mass destruction.

He went on to talk in some detail about the value of research on the bomb, and concluded that the testing program would continue. That Eisenhower attributed to Stevenson views he had not expressed—that Stevenson was talking not about atomic research but about peace—either escaped the attention of the press or was not thought important. At any rate, Eisenhower did not again comment on stopping H-bomb tests until the proposal became a major campaign issue many months later. Stevenson, for his part, was deeply engaged until September in campaigning against fellow Democrats for the nomination, and the H-bomb was not an issue in that contest. Indeed, his chief opponent, Senator Estes Kefauver, at once endorsed Stevenson's position.

But as the election campaign got under way at the end of September, Stevenson was again in a position to offer alternatives to the whole country and to plead for support of his program. He gave his H-bomb policy a high priority and elaborated on it vigorously whenever opportunity occurred. By the end of September Eisenhower was calling it a "theatrical national gesture," and other Republican speakers were treating it with ridicule. However, a growing number of eminent scientists began to speak out in support of Stevenson, emphasizing the dangers risked from the dissemination into the atmosphere of the poison strontium 90 in the fallout from H-bomb explosions. Other scientists, including Atomic Energy Commissioner Willard Libby, came out in support of Eisenhower, arguing that the danger of strontium 90 was grossly exaggerated, and that a greater risk would be run by not continuing the tests.

As in the case of the draft, the press overwhelmingly rejected Stevenson's views. And again many of his political supporters warned him that he was engaged in a losing cause. His answer was not only to persist but to develop his position more and more fully. Over and over again he told his audiences that he would not be sidetracked from the great issues. A presidential election campaign, he said, should be a "great dialogue" between the parties, whose aim is not just winning votes but educating the people. He would take as much political advantage as he could by underscoring Eisenhower's unwillingness to be drawn into a debate, and he would smoke out his opponent if he could.

On October 15 Stevenson devoted an entire nationwide television program to the H-bomb question. He began by recalling his own experience of seeing the horrors of war in Italy thirteen years before—"It was painfully clear, there in the Liri Valley, that civilization could not survive another world war. And that fact became even more clear on the day the first atomic bomb exploded over Hiroshima." He spoke of his resolve to work for peace, his service at the United Nations, his entry into politics:

> And now, thirteen years after that decision in Italy, I come before you to talk a little about the cause which means more to all of us than anything else—the cause of peace.
>
> We are caught up today, along with the rest of the world, in an arms race which threatens mankind with stark, merciless, bleak catastrophe.
>
> . . . In this nuclear age peace is no longer merely a visionary ideal, it has become an urgent and practical necessity.

But the necessity for peace did not preclude the necessity for armed strength. Disarmament must be universal, not unilateral:

> But nations have become so accustomed to living in the dark that it is not easy for them to learn to live in the light. And all our efforts to work out any safe, reliable, effective system of inspection to prevent evasion of arms agreements have been blocked by the Soviet rulers. They

won't agree to let us inspect them; we cannot agree to dis-
arm unless we can inspect them. And the matter has been
deadlocked there for eleven years.

Yet if we are going to make any progress we must find
means of breaking out of this deadlock.

The circumstances were imperative; his proposals were his
answer to the immediate need:

It was with this hard, urgent need in mind that I proposed
last spring that all countries concerned halt further tests
of large-size nuclear weapons—what we usually call H-
bombs. And I proposed that the United States take the
lead in establishing this world policy.

It was politically shrewd to argue here that the proposal was
not political:

I deliberately chose to make this proposal as far removed as possible from the political arena. It was made
four months before the party conventions. It was made
to the American Society of Newspaper Editors. It was
made without criticism of the present administration's
policy for H-bomb development.

Others—and not I—have chosen to make this proposal
for peace a political issue. But I think this is good. After
all, the issue is mankind's survival, and man should debate
it, fully, openly, and in democracy's established processes.

He now offered three compelling reasons why the attempt to
stop the tests should be made:

First, the H-bomb is already so powerful that a single
bomb could destroy the largest city in the world. . . .

Second, the testing of an H-bomb anywhere can be quickly detected. . . .

Third, these tests themselves may cause the human race unmeasured damage. . . .

Stevenson went on to enlarge on the dangers of strontium 90 in harsh and frightening language. Then he rounded out his argument by answering the two chief objections President Eisenhower and Vice President Nixon had made, when they were willing to deal with the issue at all:

It is said that it [his proposal] does not provide for "proper international safeguards." This misses the point, for, as scientists have long explained and the President himself has acknowledged, we can detect any large explosion anywhere.

It is said that other countries might get the jump on us. The President implied that he would stop our research while others would continue theirs. But I have made no such suggestion, and obviously we should not stop our research. We should prepare ourselves so that, if another country violated the agreement, we could promptly resume our testing program. And I am informed that we could be in a position to do so—if we have to—within not more than eight weeks.

With his policy thus set forth simply and clearly, and the chief objections answered as effectively as controversial objections can be, Stevenson now pleaded his case on grounds of moral influence throughout the world:

And, finally, I say that America should take the initiative; that it will reassure millions all around the globe who are troubled by our rigidity, our reliance on nuclear weapons, and our concepts of massive retaliation, if mighty, magnanimous America spoke up for the rescue of man from the elemental fire which we have kindled. . . .

The search for peace will not end, it will begin, with the halting of these tests.

What we will accomplish is a new beginning, and the world needs nothing so much as a new beginning.

People everywhere are waiting for the United States to take once more the leadership for peace and civilization.

We must regain the moral respect we once had and which our stubborn, self-righteous rigidity has nearly lost.

Finally, I say to you that leaders must lead; that where the issue is of such magnitude, I have no right to stand silent; I owe it to you to express my views, whatever the consequences.

I repeat: this step can be taken. We can break the deadlock. We can make a fresh start. We can put the world on a new path to peace.

At Cincinnati, a few days later, Stevenson offered the same argument and the same plea in a full-dress speech on foreign policy. At every stop of his campaign trip and in every talk, he stressed his H-bomb proposals.

If his own efforts were insufficient to dramatize the issue in the public mind, he was now "helped" by the intervention of Soviet Premier Bulganin. On October 21 Bulganin wrote a letter to President Eisenhower which contained the following explosive words:

. . . We fully share the opinion recently expressed by certain prominent public figures in the United States concerning the necessity and the possibility of concluding an agreement on the matter of prohibiting atomic weapon tests and concerning the positive influence this would have on the entire international situation. . . . As far as the Soviet Government is concerned, it is prepared to conclude an agreement with the United States of America immediately for discontinuing atomic tests. We proceed, of course, on the basis of the assumption that other states having atomic weapons at their disposal will likewise adhere to such an agreement.

Stevenson had asserted his willingness to take the "consequences" of his forthright declaration of policy. Here they were—and with vicious emphasis. Had the Soviet government, beginning to believe that Stevenson might be elected, wished to obviate the moral and political effect his H-bomb policy would have on world opinion? Or, on the contrary, had the Soviets hoped merely to confuse the American public? Did the Soviets hope, by associating Stevenson's proposal with American distrust of Russia and communism, to destroy Stevenson politically and be rid of his move to halt the bomb tests? Or was the letter perchance written in good faith? Whether one of these or some other motivation underlay Bulganin's letter, it is certain that the result was to produce a moment of national unity in the United States. The President angrily denounced Bulganin's presumptuous interference in the internal affairs of the nation:

. . . The sending of your note in the midst of a national election campaign of which you take cognizance, expressing your support of the opinions of "certain prominent public figures in the United States" constitutes an interference by a foreign nation in our internal affairs of a kind which, if indulged in by an Ambassador, would lead to his being declared persona non grata in accordance with long established custom.

Stevenson immediately gave the President his full support in these sentiments, as did the American press, in unanimous chorus.

But Eisenhower went on to reject in categorical fashion the idea of stopping the tests:

> . . . To be effective, and not simply a mirage, all these plans require systems of inspection and control, both of which your government has steadfastly refused to accept.

This pronouncement moved Stevenson to maintain his policy as he had done in April, despite the inevitable adverse reaction he could expect from newspapers:

> I share fully President Eisenhower's resentment at the manner and timing of Premier Bulganin's interference in the political affairs of the United States. . . . But the real issue is not Mr. Bulganin's manners or Russian views about American politics. The real issue is what we are going to do to save the world from hydrogen disaster. Viewed from the standpoint, not of politics, but of peace, I think the President's reply is unfortunate.

He sought to turn the Bulganin overture to positive purposes:

> There are two possibilities. One is that Bulganin's offer is made for propaganda purposes only. . . . If that is true, it should be exposed for all the world to see. The other possibility is that the Russian offer, ill-timed as it is, reflects an opportunity to move ahead now toward a stop to the further explosion of hydrogen bombs. In either event, there seems to me only one course to follow. That is to pursue this opening immediately and all the way.

It was a typical irony of Stevenson's political career that the Eisenhower administration, despite its denunciation of both Stevenson and Bulganin, did move, immediately after its triumphant return to power, "to pursue this opening."

The press, hostile in its reaction to Stevenson's H-bomb policy, nevertheless, like Bulganin, helped him to keep his views near the center of public attention, until finally Eisenhower and his advisers felt obliged to make a serious reply and defense of the testing program.

On October 23 the White House issued a detailed statement of the government's "official" position on the whole matter of H-bomb production, research, and testing. The press quickly dubbed it a "White Paper," though its quality was that of a campaign document. Among the many points listed and defended the most salient were these:

(1) Peace is a prerequisite to ending bomb tests.

(2) American policy is to maintain "quality and quantity" of military weapons to deter aggression until international trust can be established.

(3) The Eisenhower administration has a clear record of promoting peaceful uses of atomic energy.

(4) Disarmament policy must achieve "effective safeguards and controls" in any concrete program, that is, an inspection system which the Soviet has always refused.

(5) Because of the Soviet's intransigence the United States must continue testing and stockpiling nuclear weapons.

(6) Such testing is not a peril to human beings, according to the National Academy of Sciences, but beneficial in producing weapons with less fallout.

(7) Small as well as large weapons have fallout, so that ending H-bomb tests would not end fallout.

(8) A voluntary agreement could not be self-enforcing since the vast Soviet land mass makes detection uncertain.

(9) We would lose our lead if we stopped testing and

then had to resume because of Soviet violations, since a year would be required to put the American program back on a testing basis.

The White Paper was careful to stress a "long association" of the President with the development of atomic weapons. A few days later Eisenhower re-emphasized these views, assuring the people that America was in "instant readiness to lay aside all nuclear weapons—including their testing —when, but only when, we have sure safeguards that others will do exactly the same."

With campaign time running out, Stevenson set to work on a "program paper" of his own in answer to Eisenhower's. He had the voluntary assistance of several eminent nuclear scientists, while other scientists took to the air and the press to make their own replies to Eisenhower's claim that testing was not deleterious to human beings. When Stevenson's paper appeared, however, it was October 29, and war had broken out in Suez. His carefully prepared statement passed almost unnoticed.

In perspective the Stevenson paper warrants serious attention, for the H-bomb, quite indifferent to the outcome of the American election or the armistice in Suez, remained to haunt the world. Stevenson's chief points were the following:

(1) The President's argument is one of "defeatism," since he rejects all ideas for stopping tests without inspection.

(2) The Stevenson proposal would not weaken but would strengthen American defenses, because to freeze the H-bomb race "at the present level" would preserve the American lead.

(3) The President is "insensitive" to the dangers from fallout, since there is a great body of scientific opinion attesting to this danger, including other portions of the same

report of the National Academy of Sciences quoted by the President.

(4) Further, the fallout from one H-bomb is greater than from one thousand atomic bombs of the Hiroshima type.

(5) The President is deliberately misleading in regard to detection, since Stevenson's proposal deals only with H-bombs whose explosion, everyone agrees, can be detected anywhere.

(6) The President's apparent reliance on nuclear weapons "cancels terror with terror," since the Russians will shortly have as great a destructive capacity as does the United States.

(7) The President ignores the moral and political advantages of the Stevenson proposal in influencing non-nuclear nations to give their allegiance to democracy rather than communism, and rejects a great opportunity to reassert American world leadership.

Stevenson concluded his statement by quoting Albert Einstein, father of atomic energy, who replied to the question as to what weapons would be used in World War III, "I don't know what terrible weapons will be used in World War III. But I do know the weapons which will be used in World War IV—they will be sticks and stones." To his statement Stevenson appended a detailed point-by-point analysis of the Eisenhower White Paper.*

Within two weeks of the 1956 election the disarmament question was formally reopened by a message from the Soviet Union to the United States, Britain, France, and India, suggesting that these powers join with the Soviet Union in a

_____

* The text of the Stevenson H-bomb paper is published in *The New America,* New York: Harper & Bros., 1957. The full text of the Eisenhower paper was published in the *New York Times* of October 24, 1956.

new disarmament conference. President Eisenhower replied
in a letter released on January 2, 1957:

> You suggest further meetings of heads of government.
> I could agree to a meeting whenever circumstances would
> make it seem likely to accomplish a significant result. But,
> in my opinion, deliberations within the framework of the
> United Nations seem most likely to produce a step for-
> ward in the highly complicated matter of disarmament.
> Accordingly, the United States will make further propo-
> sals there.

Thereafter for many months the world's attention and hopes
were centered on United Nations discussions, first in New
York and later at a special disarmament conference in
London. It would be tedious to review all of the proposals
and counter-proposals made in these meetings, all of which
came to little or nothing. But it is illuminating to trace the
development of American policy on the testing of nuclear
weapons.

It was Adlai Stevenson, private citizen once more, who
again precipitated the discussion of testing. In an article
in *Look* magazine (issue of February 5), he restated his
position in careful detail and with great force, with the additional persuasive note that in defeat he could not hope to
make political capital by continuing his appeal. And he added
a new element:

> . . . I confess I had not anticipated the curious ferocity
> of the Republican responses. There was, indeed, reason
> to believe that the National Security Council itself *between
> September 5 and September 19* had voted "unanimously"
> in favor of a similar super-bomb proposal; but this deci-
> sion had been set aside for obviously political reasons, and

my suggestion for *strengthening* our position morally and physically in the world was grievously distorted and assailed by Republican campaign orators as a proposal to *weaken* our defenses.*

Immediately the President was asked to react:

> *Question:* Stevenson, in *Look* magazine articles, stated that the National Security Council, during the campaign in 1956 voted unanimously to halt H-bomb tests. Is that a fact?
>
> *Answer:* Of course, as you know, as far as Intelligence was concerned, he was briefed all during the campaign, and I don't know exactly what information you might say auxiliary to Intelligence, may have been given him. Now I can't either deny or affirm what he says because you know I make it a practice never to give a hint of what is a National Security Council conviction. But I can point out one thing, and I should: the National Security Council is set up to do one thing—advise the President. I make the decisions and there is no use trying to put any responsibility on the National Security Council—it's mine.

While Eisenhower's answer certainly hedged, leaving abundant room to suppose that the National Security Council had indeed taken a position similar to Stevenson's, it was nevertheless clear that Eisenhower himself remained categorically opposed to suspending or halting H-bomb tests, hence American proposals at the disarmament conferences presumably would not involve an end to testing. This quickly proved to be the case. The first 1957 American proposals were in all essential respects the same as in previous years— no disarmament or nuclear limitations without "adequate

* The President is chairman of the National Security Council.

safeguards," and "open skies" inspection systems as between the U.S.S.R. and the United States.

In March President Eisenhower met in Bermuda with the new British Prime Minister, Harold Macmillan, to discuss Anglo-American relations and disarmament. It was evident from their joint communiqué, March 25, that the problem of suspending H-bomb tests was given careful attention. Though again the decision was in the negative, it was more cautiously worded:

> Over the past months our Governments have considered various proposed methods of limiting tests. We have now concluded together that, in the absence of more general nuclear control agreements of the kind which we have been and are seeking, a test limitation agreement could not today be effectively enforced for technical reasons, nor could breaches of it be surely detected. We believe, nevertheless, that even before a general agreement is reached self-imposed restraint can and should be exercised by nations which conduct tests.
>
> Therefore, on behalf of our two Governments we declare our intention to continue to conduct nuclear tests only in such manner as will keep world radiation from rising to more than a small fraction of the levels which might be hazardous. We look to the Soviet Union to exercise a similar restraint.

The next day, in response to the Eisenhower–Macmillan communiqué, the Soviet government took the initiative by asking for a temporary ban on all nuclear tests. The Western Allies were pictured as obstructionists on the road to peace.

For several weeks Harold Stassen, American representative at the disarmament conference, devoted his efforts to securing agreements to inspection systems and safeguards against surprise attack as preconditions to an agreement to suspend nuclear weapons testing. Inspection, as it had been in all past conferences, again proved nonnegotiable. By April 23 Secretary Dulles was announcing that since no "sound" agreement had been reached, the United States would continue testing unless there was "new scientific information" that such tests would be perilous to the world's health. The Atomic Energy Commission presently supported Mr. Dulles by a statement that nuclear tests "at present levels" created no danger to mankind.

Stevenson, appearing May 5 on "Meet the Press," a nationwide television program, took the occasion to restate his views. Showing no confidence in the position of the Atomic Energy Commission, he called for an end to "contamination without representation." By this time many governments throughout the world, as well as Pope Pius, had expressed concern at the continuance of hydrogen bomb testing, and on June 1 the *New York Times* reported that the United States was giving serious consideration to a testing "limitation" program as a "first step" to a general accord on disarmament. But at his press conference of June 5 President Eisenhower repeated his insistence that testing could be stopped only when there was a guarantee of safeguards and inspection. He referred back to the report of the National Academy of Science issued a year before:

. . . That is the authoritative document by which I act up to this moment, because there has been no change that I know of. Now, on the other hand, here is a field where scientists disagree. Incidentally, I notice that in many instances scientists that seem to be out of their own field of competence are getting into this argument, and it looks like almost an organized affair.

I am concerned just as much as I am of this fallout. [*sic*]

I am concerned with the defense of the United States. I have tried, and this Government has tried, to make the abolition of tests a part of a general system of disarmament, controlled and inspected disarmament.

If we can do that we will be glad enough and very quick to stop the tests. But we have the job of protecting the country.

Nevertheless, the American position was being slowly altered. Two weeks later, on June 19, the President was asked whether his insistence—that there could be no stopping of tests without a firm agreement that nuclear weapons would never be used in war—was to be applied to the Soviet proposal of a temporary suspension:

> *Answer:* Oh no. No, no. I would be perfectly delighted to make some satisfactory arrangements for temporary suspension of tests while we could determine whether we couldn't make some agreements that would allow it to be a permanent arrangement.
> *Question:* Would this be a suspension without conditions?
> *Answer:* No, I didn't say suspension without any conditions at all. If I gave that impression I made a very great error. It would have to be a suspension under such a method that we both knew exactly what we were doing, and then, as I say, using that interval to work out something in which we could have real confidence.

In London on July 2 Stassen countered a Soviet proposal for a two-year suspension of tests with a proposal to suspend tests for ten months, together with an agreement to ban the use of nuclear weapons in war. The Soviets rejected the proposal. On July 11 Stassen repeated his proposal and invited the

Soviet Union to choose between short periods of test suspension, preferably ten months, or no suspension at all. The Soviets remained adamant and the conference remained deadlocked for about six weeks. Finally, on August 21, President Eisenhower agreed to suspend the testing of nuclear weapons for two years if the Soviet Union would agree that during that period there would begin a permanent cessation of the production of fissionable materials for military purposes. In this limited and crippled form Stevenson's "national theatrical gesture," as Eisenhower had called it, at last became American policy.

After nearly eight months of fruitless discussion and negotiation, the United States was officially committed to a policy of stopping the testing of H-bombs. But by its insistence on including all nuclear tests the Eisenhower policy missed the most effective element of Stevenson's proposal, that a beginning should be made with the H-bomb, whose explosions are the chief source of poisonous fallout, yet can readily be detected anywhere in the world. No one could say whether such a policy would at any time have been acceptable to the Soviet Union. But in any case it was now too late. The Soviet rejected all American proposals on the ground that they were too thickly encased in conditions. The London conference broke up, and Stassen returned to Washington emptyhanded. There was even some reason to suppose that Stassen had exceeded his instructions.

Meanwhile, on August 26, the Joint Congressional Committee on Atomic Energy published a report of its spring hearing on radioactive fallout. Included was a report of a meeting of experts held in Washington on July 29 at which there was cautious agreement that if the nuclear powers continued their testing at a rate equivalent to the rate of the past twelve years, a danger from fallout would develop. Among the scientists concurring were the head of the Atomic Energy

Commission's division of biology and medicine, and Atomic Energy Commissioner Willard F. Libby, who had earlier minimized fallout dangers on behalf of the AEC.

Throughout the many months of the disarmament conferences and in the continuing scientific debate, the dangers from fallout seemed always to be considered on the basis of "present levels" of testing. Yet in 1957 Great Britain joined the hydrogen powers and began a large-scale testing program. And there was nothing to prevent other nations from following suit, as Stevenson had long before pointed out. But if the American, British, and Soviet governments were not yet ready to bring an end to man's gradual self-destruction in the name of self-defense, the evidence was mounting that other leaders of opinion, and plain people everywhere, were. A letter by the revered Dr. Albert Schweitzer, calling in the name of human dignity for an end to the tests, was reprinted in newspapers all over the world. Many an American editor who had denounced Stevenson in 1956 now raised his voice with Schweitzer's. An end to the testing of nuclear weapons seems an inevitable outcome of the human choice between peace and destruction. But the American decision, in August, 1957, to approach the policy advocated by Stevenson more than a year before, came too late to bring with it the moral and political leadership he had envisioned for the United States.

When the opportunity presented itself during his NATO mission late in 1957, Stevenson reopened the whole disarmament question—this time from a post perhaps strategically more promising than that of presidential candidate or opposition leader. At any rate there is reason to suppose that, events having marched and campaigns receded into the background, his views, with massive backing from world opinion, had some real influence in the shaping of American disarmament policy. He set forth these views, November 4, 1957, in a private memorandum for the Secretary of State:

I *repeat* that more flexibility and more initiative toward disarmament, at least nuclear disarmament, is imperative. We must be ready for new Russian proposals which may be extremely effective as propaganda. I have often said that suspension of nuclear testing with suitable monitoring posts to safeguard against violation should not be made conditional on cessation of production of nuclear materials. It would be an important first step; it would break the deadlock and halt or slow down the dangerous race which at best leads only to stalemate and a balance of terror.

At Paris the American delegation announced no change of policy, and another opportunuity was wasted. As late as January 13, 1958, Eisenhower was writing to Premier Bulganin again rejecting test suspensions without suspension of production. "You renew," Eisenhower wrote, "the oft-repeated proposal that the United States, the United Kingdom, and the Soviet Union should cease for two or three years to test nuclear weapons . . . [but] you defer to the indefinite future any measures to stop production of such weapons." Yet three months later when Khrushchev, newly become Russian Premier, notified Eisenhower of the Soviet intention to stop testing, the American position underwent a significant modification. In his reply to Khrushchev of April 8, President Eisenhower announced that the "United States is also prepared in advance of agreement upon any one or more . . . propositions, to work with the Soviet Union, and others as appropriate, on the technical problems involved in international controls." Soviet acceptance of this significant proposal led to conferences of technical experts in the summer of 1958 which reached agreement on both the feasibility and the methods of control. The agreement of the scientists led in turn to renewed disarmament negotiations at Geneva, beginning on October 31, 1958.

The agreement of the scientists marked a measure of progress, but hopes were once more dashed when the Geneva conference bogged down in familiar rigidities and familiar "conditions" attached to proposals. The American position became more flexible on the matter of production of military nuclear materials, but adamant on inspection conditions, while the Soviet Union introduced a new complication—the right to "veto" test inspections.

In April, 1959, the American position again underwent a significant change. On April 20, President Eisenhower wrote to Khrushchev proposing a new basis for reopening the Geneva negotiations. By way of preface Eisenhower's letter contained a paragraph which well illustrates how far he had come since October, 1956:

> The United States strongly seeks a lasting agreement for the discontinuance of nuclear weapons tests. We believe that this would be an important step toward reduction of international tensions and would open the way to further agreement on substantial measures of disarmament.

The resemblances to Stevenson's oft-used language were surely coincidental, but the reversal of views expressed by the President was full recognition that his opponent had been right and had, indeed, shown the way that would in any case have to be pursued. It was probably small comfort to Stevenson that newspaper columnists like Drew Pearson and Roscoe Drummond rather unrealistically chided Eisenhower for his ungracious failure to acknowledge that his rival had the better of the argument, or, indeed, that the departed Harold Stassen had been right and Dulles wrong. But Eisenhower now went much further toward Stevenson's view. Dropping entirely the matter of stopping production of materials for nuclear warfare, he even accepted on a temporary basis the Soviet "veto" thesis in order to make a new offer:

> In my view, these negotiations must not be permitted to fail. If indeed the Soviet Union insists on the veto on the fact-finding activities of the control system with regard to possible underground detonations, I believe that there is a way in which we can hold fast to the progress already made in these negotiations and no longer delay in putting into effect the agreements which are within our grasp. Could we not, Mr. Chairman, put the agreement into effect in phases beginning with a prohibition of nuclear weapons tests in the atmosphere?

This suggestion, the President argued, could be carried out without the "automatic on-site inspection which has created the major stumbling block in the negotiations so far." Again he expressed the hope that the Soviets would be prepared to go further and negotiate a comprehensive agreement. But failing that, the President reiterated his hope that they would agree to this new minimal step:

> If we could agree to such initial implementation of the first —and I might add the most important—phase of a test suspension agreement, our negotiators could continue to explore with new hope the political and technical problems involved in extending the agreement as quickly as possible to cover all nuclear weapons testing. Meanwhile, fear of unrestricted resumption of nuclear weapons testing with attendant additions to levels of radio-activity would be allayed, and we would be gaining practical experience and confidence in the operation of an international control system.

Thus Eisenhower at last completed the reversal of his views. At the outset, in 1956, he had asserted that weapons testing was essential for national defense, that, indeed, it was folly to propose their suspension, and that fallout was not dangerous. In the face of mounting American and world opinion he had agreed in 1957 to suspend tests, but only if a foolproof inspection system were adopted and a simultaneous agreement were entered into to cease the production of nuclear war materials. The next stage, 1958, was to agree to suspension for a limited time with only an agreement on inspection. Finally, in 1959, he offered to begin a "phased" program of test suspension which would involve in the first stage not even an on-site inspection system. Eisenhower's first phase was different from that which Stevenson had proposed in 1956, but the point was the same—to make a beginning, because something must be done to break the deadlock, because a beginning would bring hope of better things, and because the fear of fallout would be alleviated. Thus the President once again found himself following the leadership of his twice-vanquished opponent. And irony again prevailed in the high politics of the United States in the 1950's.

## CHAPTER VII

# "Titular Leader" Again,

# 1957–59

I

Shortly after his defeat in 1956, Adlai Stevenson made it clear that he would not again seek the Presidency. Though at no time did he assert that he would not accept the Democratic nomination if it were offered, he sought to persuade his friends and supporters that he wished to step aside in 1960 in favor of a younger man. But he had no wish to escape the responsibilities of party leadership before that time came. As he had done in 1953, he undertook a rigorous schedule of party dinners to raise money for the party campaign deficits, and sought ways in which to offer effective opposition to the administration.

Of the two innovations in party organization and procedure for which Stevenson was in whole or in part responsible— open convention choice of a vice-presidential candidate and the Democratic Advisory Council—there is no doubt that the latter is the more important. Stevenson's decision to throw open the nomination for the Vice-Presidency at the 1956 convention was severely criticized by a number of politicians at the time, and in 1960 Senator John F. Kennedy reverted to earlier custom in his personal selection of Senator

Lyndon Johnson to be his running mate. It is certainly by no means clear which is the better method of choosing vice-presidential candidates, and the Stevenson innovation deserves careful study by political scientists and by leaders in both parties. But the advisory council idea seems likely to persist, and the Democratic Advisory Council itself, the first American "shadow cabinet," gave the party a new and effective national voice unparalleled for a party out of executive power. By 1959, indeed, the Republican party had begun to experiment with a similar adjunct to its national committee.

After his crushing 1956 defeat, Stevenson could not hope to exert more than formal political leadership within the Democratic party. While it remained possible that he would be nominated a third time, he had no longer any claim upon the party. In practice he had to share his leadership of the opposition. Under similar circumstances in earlier years, opposition from the "shadow" executive has typically almost vanished, while the congressional wing of the party asserts its leadership. Thus Thomas E. Dewey could remain an influential figure after 1948 but had to surrender his commanding position and share leadership of the Republican party with Senator Robert Taft and others.

At the start of the new Eisenhower term in 1957, the moderate Democratic congressional leadership of Speaker of the House Sam Rayburn and Senate Majority Leader Lyndon Johnson made it clear that they expected to assume control of the party and to follow a line of polite collaboration with the Republican administration. But this time the non-congressional leadership was unwilling to go along. Stevenson and National Chairman Paul Butler were convinced that there should be a serious effort to organize an effective means of articulating opposition to those Republican policies which representative Democrats thought ill-advised, and to formulate

positions which Democrats could advance on the chief continuing issues before the nation. Both Stevenson and Butler
hoped that the congressional leadership would join in the
effort but were prepared to go ahead in any case. While congressional co-operation would be desirable, the point was to
organize a voice for the national party, or "shadow administration," which members of Congress, chosen to represent
local or state constituencies, were not equipped to do.

Chairman Butler, acting under a resolution of the Democratic National Committee, established a Democratic Advisory Council empowered to formulate party policy for the
National Committee between conventions. Besides Stevenson, Butler appointed to the council former President Truman, Governor Averell Harriman of New York, Governor
G. Mennen Williams of Michigan, David Lawrence of Pennsylvania, Jacob Arvey of Illinois, Paul Ziffren of California,
among others; and such congressional figures as Senators Estes
Kefauver, Hubert Humphrey, and John F. Kennedy, who
did not belong to the congressional leadership but had
achieved national stature in their own right. Speaker Rayburn
and Senator Johnson were invited but refused to join. The
Advisory Council began to function during the winter of
1957 and has grown steadily in power and influence ever
since. As it grew, it enlarged its scope to include several
advisory panels appointed to provide the council with draft
statements and papers on such specialized subjects as foreign
policy, economic policy, agricultural policy, civil rights, and
others. At the Democratic National Convention in 1960 a
resolution was adopted to authorize an advisory council
whenever needed. John Kennedy's victory, of course, brought
a cessation of the council's activities, but not of its availability for future service.

While Paul Butler served as chairman of the Advisory
Council from 1957 to 1960, and deserves a major share of

the credit for its institutional development, it is nevertheless clear from the council's many statements and position papers that Stevenson was the leading intellectual force among its members. As titular leader Stevenson could speak for the party only, as it were, ceremonially. But through the Advisory Council he could speak as *de facto* leader of an opposition that could effectively challenge the administration and command a national hearing for its proposed public policies. The measure of Stevenson's influence in the council is the consistency with which its statements and papers followed his expressed views. At the same time it is certain that Stevenson's positions were sharpened and sometimes modified by the critical exchanges of the council sessions and the skillful preparatory work of the council's staff under Charles Tyroler. Thus Stevenson's personal leadership of the party was transformed into a kind of collective leadership in which he continued to play a central role.

## II

Aside from his work in the Advisory Council, Stevenson still found himself in a position to exert personal influence to a greater extent, perhaps, than any other private citizen of his time. Requests for him to speak before organizations and meetings, large and small, flowed into his Chicago law office at the rate of five to twenty-five a day. And though he deliberately reduced his public appearances to as few as he possibly could, he nevertheless continued to make a good many speeches, and his appearances always commanded national attention.

Though there were some disadvantages in his new role—owing to his diminishing political influence in his party—Stevenson found some compensatory advantages. He felt much less pressure from friends and advisers to say this or that, to take a certain position, or to concentrate on certain subjects which might not be agreeable to his interests. As a non-candidate, in short, he was considerably freer to pick his own themes and to develop them as he thought best.

Early in 1958, partly as the result of his frustrating experience as adviser on NATO in the State Department, Stevenson began to develop an emphasis on foreign economic policy which, like so many other vital matters he had brought to the forefront, presently assumed a new dimension of importance both in and out of government, at least partly because of his leadership.

In the State Department Stevenson had found administration leaders deeply committed to the "mutual security" idea. The Eisenhower–Dulles policy of contracting as many defensive alliances as possible to shore up the free world against communism had, of course, originated in the Truman era, and its value had been tested in the Korean War. But with that war receding into history and the balance of nuclear terror more and more controlling world political relations, many people began to raise questions about the continuing emphasis on the military aspect of American relations with the free world. As early as 1949 President Truman, in "point four" of his annual message, had suggested a future American involvement in the underdeveloped nations of the world which would have no military significance. American dollars in small amounts and American skills in substantial amounts would be exported to help peoples engaged in the "revolution of rising expectations" (Stevenson's later phrase) to raise their standards of living in food, education, health, sanitation, and similar matters. But the small beginnings of technical assistance were nearly lost in the massive programs of mutual security after 1950. By 1957 it seemed to Democrats like Chester Bowles and Republicans like Paul Hoffman, as well as to Stevenson himself, that American policy ought to undergo a major shift of emphasis, away from military assistance with economic aid mainly as support for military programs, and toward economic assistance for its own sake—technical assistance, capital transfers, and substantial loans for long-term development projects in many parts of the world.

In his 1956 campaign Stevenson had made an effort to get attention for such needs as these. Indeed, he had been talking about economic development since his world tour of 1953. But prevailing fears of overt Communist aggression tended to throw all discussion of development programs into the military perspective. Events in the fall of 1957 shifted the focus of attention as thoughtful attempts at persuasion had failed to do. After sputnik it became clear that the primary Russian challenge was not military but scientific, economic, and cultural. It was the hope that the administration would shift its policy accordingly that led Stevenson to attempt collaboration with Dulles and Eisenhower. When the administration nevertheless insisted upon a continuation of its emphasis upon military defense, offering in NATO no more than pious words in support of a policy for foreign aid geared to development of the underdeveloped world, Stevenson chose thereafter to give the major emphasis of his public activity to analyzing the problem and pleading for a massive long-range development program. During the spring months of 1958 he took every occasion to hammer out a policy and make it understood. The keynotes of this new campaign were struck at the Roosevelt Day dinner in New York on February 1:

The Communists have moved quickly in another direction [i.e. different from the technological] . . . the political, economic and psychological penetration of the backward, under-developed areas where freedom from external control is new, economies are weak, governments unstable, and the people yearn for material improvement.

I am confident that the Russian–Chinese bid by aid, trade and propaganda for influence and allegiance in these areas is far more dangerous than Soviet missiles or manpower. For these areas of Asia, the Middle East, Africa and Latin America with their vast populations and resources are decisive in the world struggle. We need them and they need us. To help these countries develop so they can better produce what they sell, so that they can better buy what we sell, and so that they can better preserve their independence, is the best investment we can make.

We have been losing the cold war in these areas as well as in weapons. To measure Communist success in political penetration by economic aid and subversion we have only to look around—at Egypt, Syria, and Southeast Asia. If this goes on, the consequence will be progressive isolation for the United States.

For these reasons, Stevenson went on to support the administration's program for foreign aid, even though he thought it much less than adequate, and to attack the Democratic Congress for its tendency to look upon aid programs and budgets as easier to cut than any others, on the favorite ground of "money down a rat hole."

Stevenson then turned to a drawing of the issue between a policy chiefly military and a policy of response to the needs of peoples:

The Russians saw that Point Four, economic assistance, was what these countries wanted and went for it in a big way. But our emphasis has been largely military—so much so that the world has begun to suppose that America thinks only in terms of military solutions. Our diplomacy has been addressed primarily to the promotion of military coalitions around the Communist periphery. Ninety percent of our foreign aid, so called, has been allocated on military grounds. Meanwhile, nations which were unwilling to join us in military pacts have had relatively little economic assistance. This has played directly into Moscow's hands. And I am not sure how effective our huge military assistance programs have been.

The Communists, he declared, had outwitted us consistently and assumed the role the United States ought, by character and inheritance, to be playing:

> . . . the Communist bloc has rapidly increased its long-term, low interest credits to strategic countries for economic aid. They have a clear-cut strategy for dealing with the surging nationalism of the new nations—to keep it aimed forever against their old colonial masters, and by skillful penetration and subversion to gradually chain these decisive regions to their victorious chariot.

Communist operations in these underdeveloped areas ought to contrast sharply with American in character and purpose, since American motives are wholly different:

> As for us, what do we want of the people who live in these regions? We want them to remain independent; to modernize their ancient societies; to develop their countries economically; to take a role of increasing responsibility on the world scene; to succeed for their sake which is also ours.
>
> And we should know they can preserve freedom and contribute to world order only by fulfilling the aspirations of their peoples for human dignity and a tolerable standard of living. For that they need help—loans and credits mostly

—and ultimately they are likely to take it where they can get it.

As in so many other matters of high policy, it was not long before events confirmed Stevenson's forecast. By 1960 the boiling cauldron of the Congo and the violent anti-Americanism of Castro's Cuba attested to the folly of trying to meet revolutionary thrust with programs geared chiefly to mutual security against communism.

Again, as he had done during his State Department days in previous months, Stevenson called for co-ordinating the efforts of Western democracies to assist the underdeveloped areas in a manner similar to the co-ordination already achieved in military matters. And he made a quick suggestion which he was later to develop in detail:

> Perhaps it is not beyond the genius of Western capitalism even to help stabilize commodity prices. The present decline, according to the UN figure, has cost the primary producers, who are mainly uncommitted, $600 to $700 million in export income in 1957, which wipes out our economic aid program, and then some!

On February 25, 1958 at Washington, Stevenson made one of the main addresses at the National Conference on Foreign Aspects of U.S. National Security which was called to demonstrate bipartisan support for the administration's foreign aid program. Stevenson gave the necessary support, but devoted most of his attention to developing his own, much more ambitious program. He summarized his proposals in six points:

> *First:* This year the full $625 million requested should be appropriated to the Economic Development Fund, and

next year the Fund should be put on a permanent basis so that it can plan investments forward for some years, not merely on a project basis but in terms of the total requirements of the receiving nation. Our emphasis should be on loans not gifts where feasible. You remember Confucius' question: "Why do you dislike me? I have never done anything to help you!"

*Second:* We must make clear our support for the Indian 5 year plan on a long-term basis and enlist other nations to do likewise, so that the Indians can proceed with reasonable confidence that they can achieve their essential objectives. India—twice the size of the whole Marshall Plan world—will need a sustained support to get over the hump. If we fail there, in the key to free Asia, our cause will suffer grievously everywhere.

*Third:* We should explore methods of increasing private capital investment, although risk capital can't and won't begin to do the job of basic development—roads, power, transportation, schools, etc. But with strong government leadership and with a steady flow of government capital, private investment will have an enlarging role to play.

*Fourth:* We must learn to use our surpluses of food and fiber as a major constructive resource in economic development, not as charity but as working capital—to enable men to divert their labor from agriculture to roads, dams, power stations and the like without creating an inflationary demand for food and clothing. . . .

*Fifth:* I think it is time we coordinated our economic affairs with our friends instead of going it alone. The whole of the industrialized free world has the same interest in seeing the underdeveloped areas make the transition to self-sustaining growth while maintaining their independence. This is a global enterprise; and it should be organized on that basis. . . .

*Sixth:* Against the background of an enlarged and stabilized American program, weaving together the great resources of the industrialized nations, giving play to private as well as government initiative, we could well invite Mr. Khrushchev to coordinate his efforts with ours if he is really interested in the economic development and political independence of these less fortunate countries. Such cooperation—and I hope we leave no stone unturned in our effort to cooperate—could avoid the waste and hazard of blackmail that results from competition in this kind of effort. And if the Soviets are not interested in joining our international effort, such an offer would at least unmask the motives behind their assistance programs.

On March 27, at another national conference in Washington, this time called to show bipartisan support for the reciprocal trade law, Stevenson added a strong plea for freer trade to his proposals for foreign economic policy:

And this is to say what ought to be self-evident: that, as no man is an island unto himself, neither is any country; not even the United States, which must import more and more of its needs and export more and more of its product to be safe and sound.

We are the last great outpost of free enterprise capitalism. Yet here we are striving to keep alive a significant phase of free enterprise in its homeland: to keep a significant phase of capitalism functioning in the country of the capitalists; to keep competition alive among the apostles of free competition!

This is odd. But do I exaggerate? Let me illustrate. Our chemical industry is the world's largest. The annual sales of only one of its units is double the size of Japan's total exports. In 1957 our chemical exports were one billion dollars greater than our imports. But this multi-billion dollar colossus wants tariff protection. . . .

The moral is chilling. Regardless of all the lessons of history, of all the counsels of experience, short-sighted domestic self-interest rather than the hard realities of international life often prevails in our public policy.

The seriousness with which he viewed economic policy and the importance he attached to freer trade within the complex of that policy may be judged from his next words:

Yet in our dangerous days this is reckless counsel. For the character of our times, the nature of our struggle for survival in a pitiless universe, will be determined largely by how we react to material events. The manner of America's dealing with foreign trade policy, for example, will illuminate the way in which we may be expected to deal with the terrible issues confronting us in other fields.

We may rise to the occasion, or we may sink to it. How we shall act must finally depend upon our resolution and our understanding of the apocalyptic day upon which the world has come.

In a series of addresses at universities in late May and early June, Stevenson brought his campaign for an enlightened foreign economic policy to an impressive climax. In what he called an "atmosphere of growing crisis" he suggested "with the utmost urgency" that a "committee of experts" be formed, comparable to the group which did the preliminary work on the Marshall Plan. His proposals for the consideration of such a committee were four:

*First, the recovery of internal momentum.* We must consider the implications of aiming at a sustained rate of growth of 5 per cent a year in place of our traditional 3 per cent. . . .

*Secondly, in the field of tariffs,* I would hope that our Commission of Experts would link the European experiment with a common market to the needs of a wider free trade area. . . .

*Third, is economic aid to underdeveloped areas.* We must rationalize and enlarge the many foreign investment programs of the industrialized nations. Six months ago I urged the Administration to take the lead in the direction of consolidating and coordinating our financial and human resources for economic development in the "have-not" nations. For maximum effectiveness, such loans and aid should be stabilized and related to long-term policies of sustained growth in the backward countries. Underpinning India's economic transformation is an immediate imperative. Germany must be brought in on the contributor's side. And, surely, to give to contributors and receivers alike a steady picture of the amounts and skills needed and available is not beyond the genius of the free world.

*Fourth, is reserves or working capital, and the convertibility of currencies into dollars. . . .* My first impression is that with larger backing from the hard currency powers, the International Monetary Fund could become an effective instrument of full convertibility. Trebling the contribution of all members would be in keeping with the modern tendency to regard all currencies, hard and soft, as part of the "cushion" to world trade.

While Stevenson's concern with these pressing economic questions was widely and deeply shared by experts and by some more thoughtful political leaders in both parties, the record of the 1950's shows that public interest in foreign economic policy was minimal. It was a difficult subject at best, and it was easy for opponents of foreign assistance programs to talk glibly about "money down a rat hole," or for enemies of free trade to complain about American workers losing their jobs because of cheap foreign competition. In the later years of the decade, especially after 1957, the Eisenhower administration made greater efforts to address the economic problem in realistic terms. Douglas Dillon, Under Secretary of State for Economic Affairs, during the years 1959 and 1960 moved, indeed, some distance along the paths Stevenson had suggested. But at the end of the decade most of United States foreign aid was still in military support. A massive attack on poverty throughout the world had yet to be mounted by the West, and leadership throughout the underdeveloped world was still confused by Soviet promises (and sometimes performance) and American uncertainty.

By the time of the 1960 presidential election there were signs that Stevenson's campaign of public education on foreign economic problems, possibilities, and imperatives had had a cumulative effect. The public appeared much better prepared to listen and discuss; the presidential candidates of both parties were less hesitant to oppose Russian aggressiveness with positive proposals for Africa, Asia, and Latin America; and, as will be noted, both party platforms called for greater emphasis on foreign aid and trade. At the United Nations in September, 1960, President Eisenhower made a strong commitment on behalf of his country and the West calling for large-scale programs of economic development in Latin America and Africa. In December, as already indicated, the OECD came into being. But the reversal of American policy was too late to have helped to avoid the chaos in the Congo or bitter violence in Cuba.

## III

In the summer of 1958, when talk of a new summit conference was beginning to be heard, Stevenson made a trip to the Soviet Union to see for himself what the Communist stronghold was like, to gauge, if he could, the temper of the Russian people and their leaders, and to observe the changes which had taken place there during the thirty years since, as a young man, he had visited the Russia of Lenin and Stalin.

Stevenson's frame of mind as he set out for the Soviet Union, and private conversations with Premier Khrushchev, may be gathered from some things he said during a remarkable three-way transatlantic radio program on March 12. In this program he had participated in a discussion of world affairs with his counterparts, the opposition leaders in the United Kingdom and France—Hugh Gaitskell and Pierre Mendes-France. The talk dealt at length with what possibilities of easing tensions might be expected from a summit conference. Stevenson adopted an attitude of caution based on his understanding of the Communist aim to move toward domination wherever possible, yet he saw in the terrifying threat of hydrogen war a new factor in world affairs which might add a dimension to Communist thinking:

> As to this forthcoming meeting, it would be a pity if our expectations were too high. There is, however, a reason for hope and optimism that we might accomplish something on a less exalted scale. Proposals have been made from time to time—as long as two years ago in this country by myself, by Mr. Gaitskell and others abroad—with respect to the discontinuance of the testing of hydrogen weapons (we believe with suitable inspection here—that

that's a necessary precaution). At least if this could be accomplished at the summit this time, this would establish the principle—the mechanics, perhaps—of an inspection system—at least look in that direction. And this would break the arms deadlock which seems to me is the most terrifying aspect of our contemporary scene. This would be a very substantial achievement, and from there we could go on to the consideration of outer space, and so on.

At the same time Stevenson made it clear that he did not think a summit meeting with Khrushchev could be expected to bring about a solution of the German problem or a full-scale reduction of tensions anywhere. Disengagement, he thought, for example, should not be attempted in "a few days amid the bright lights, the confusion, the diversions of a summit meeting." His own view was that the West should agree before the meeting and "dramatize" at the summit a "sensible progression," while seeking Soviet agreement "now as to where we can *begin.*"

Later in the same broadcast Stevenson hesitated to agree with Gaitskell that the Russian attitude was "Let's get together and make a start." He suggested that the Soviet attitude might more accurately be phrased as "Let's get together and make propaganda." On the other hand, he agreed that the attitude of the American government was too rigid, and that it was not wise to insist that there could be no summit until substantial progress had been made at lower levels. In short, he was prepared to support a summit meeting so long as only modest goals were set for it. But he was not prepared to suppose that Soviet long-range aims to communize the world had been set aside because of the need for peace.

In six weeks of intensive travel Stevenson covered some seven thousand miles of the Soviet Union from the Gulf of Finland through Siberia to Central Asia and the Chinese border. As he reported in a series of newspaper and magazine articles, he was given many opportunities to see Russian areas normally closed to foreigners. He was never in any way hindered from talking to people in all walks of life, in all sorts of stations, in many regions both urban and rural. He found a gross contradiction between the attitude of the people and the official line of the Soviet government regarding the old shibboleths of "communism vs. capitalism." When he collected his articles into a book he suggested this contradiction in the title—*Friends and Enemies*. Everywhere he found the people friendly and preoccupied with work to improve their economic and cultural life, hopeful for peace in the world, and willing enough to let the Americans, as well as other countries, live according to whatever social systems they might choose. Stevenson regretted an undercurrent of fear inspired by the ring of military bases around the Soviet Union, built under NATO and other United States treaties. He found much doubt that these bases were solely for defensive purposes. Still; he found great good will toward the Americans, lively curiosity about the United States and the West, and eagerness to co-operate and to exchange ideas.

But this long conversation with Khrushchev was of a different order. He found the Soviet leader friendly enough, but tough and unyielding on many questions. Above all, he showed no signs of relinquishing his doctrinaire communism. He was anxious to exploit the landing of American troops in Lebanon (at the invitation of that government) as aggression. He lashed out at Chiang Kai-shek as a "political corpse" with whom he would not sit down, demanded a summit conference to bring an end to "imperialist aggression," or, failing that, a summit meeting of the United Nations Assembly "with all countries participating, to condemn the aggressors and demand the withdrawal of their troops." But Stevenson concluded that such matters as these were not really central either to understanding Khrushchev or the Soviet Union. Much more significant—and encouraging—he felt, was Khrushchev's repeated admonition that "public opinion must be respected."

> I could hardly believe my ears—the Prime Minister of the Soviet dictatorship, which tolerates no criticism, was lecturing me on democracy and the sovereignty of public opinion! But I think he was saying something significant —the present leadership of the U.S.S.R. does consider opinion, for its foreign policy relies, in part at least, on persuasion rather than coercion as in Stalin's day.

Khrushchev, Stevenson reported, always "brought the talk back to his question: 'How shall we improve our political relations?' " It seemed reasonable to suppose that, in view of this attitude, some small beginnings might be made toward reducing the causes of tension, and Stevenson returned from his trip encouraged to believe that he had been right in his estimate of the situation.

But his optimism was chiefly for the long years ahead rather than in terms of present settlement:

> My happiest conclusion is that the Russian people don't want war any more than we do. The people, who suffered so horribly in the last war, don't want it for obvious reasons; the leaders because it would interrupt their great development program, and because they believe the manifest destiny of Sovietism is to inherit the earth from "decadent capitalism" anyway.
>
> Khrushchev's phrase was "We will bury you." But he did not mean that they would kill us first. On the contrary, I concluded that they would use their arms cautiously, knowing from experience that any further expansion by

force means war because we would intervene. I suspect they realize, too, that Stalin was the principal architect of NATO because he frightened us.

Stevenson remained concerned that American bases provided Khrushchev with "such a convenient peg for propaganda about America's offensive threat," but he also concluded that the Soviet leaders "may no longer believe in the inevitability of war." By 1960 his conclusion in this regard had been justified by the outbreak of an open disagreement between the Russians and the Chinese on precisely this point.

In the habits formed by institutional life Stevenson saw another encouraging sign—the "decline of fanaticism." He noted that many of the newer leaders of the Soviet Union had come up not through political ranks but through career administration as engineers or economists. "In this new managerial elite," he observed, "ideology gets more lip service than passion." Such people were not "revolutionaries trying to seize power, but realistic, practical people trying to make the Soviet system work better." They would be "easier to deal with," he thought, "as they replace the older generation of combat-Communists."

Finally, Stevenson reported his feeling that "with economic improvement and better living conditions in Russia we will have more in common and the enmity of inferiority will diminish." He was prepared to predict that with "more self-confidence" on the part of the Russians, co-operation in scientific, economic, and political matters would "broaden and reduce the tensions and divisions."

With the past perspective of his journey Stevenson saw no reason to alter the views he had entertained before he set out. Little could be expected in the short run by negotiation with Moscow, but that little was worth trying for—such as suspension of nuclear tests with inspection ("a hopeful pos-

sibility"). The real task for the United States and the West
was to "set its house in order and keep it in order, and not just
sit around bickering, postponing, and waiting for total peace
to break out." "Moscow," he said,

> will be more likely to talk seriously if the Western alliance
> is vital and viable, the residual colonial problems being
> dealt with (while the reality of Soviet imperialism becomes
> more obvious), and above all the free world making a
> concerted effort to unite the advanced and retarded areas
> in common economic enterprises.

Thus Stevenson returned not only from his long journey but
to his intellectual and political starting point—the conviction
that the future of freedom lay with economic development in
the underdeveloped world, assisted and underpinned by the
capital and talent of the West.

A year later Stevenson and Khrushchev spent a day together on the farm of Roswell Garst in Iowa. As the *New
York Times* reported their conversation, Stevenson elicited
many expressions from Khrushchev of his and the Russian
people's desire for peace. It seemed to Stevenson that the
Russian leader might "have changed his concept of America"
by seeing so much friendliness. But Stevenson still felt that a
real settlement must be a dim and distant thing: Khrushchev
would like to "reach an agreement with the United States
on terms satisfactory to the Soviet Union, but possibly not
to the United States. However, I think there are areas susceptible to negotiations."

A few days later, when Khrushchev addressed the United
Nations Assembly and offered his plan for total universal disarmament, Stevenson issued the following statement:

> Mr. Khrushchev's total disarmament proposal must be
> taken seriously. The only way to eliminate the scourge

of war is to eliminate the means of war. And Mr. Khrushchev has proposed just what we have all preached—a
disarmed world.

Whether he means what he says is the question now. We
have reason to be skeptical, but we have better reason to
study his proposal with an open mind and high hope for
progress at last towards arms control with security.

The Soviet Union knows as well as we do that in the
Nuclear Age no nation can afford unlimited war. I have
often said that a danger greater to us than war is Soviet
economic penetration around the world. So I do not dismiss Mr. Khrushchev's speech as propaganda only.

The Eisenhower administration, however, did dismiss the
Russian proposal as propaganda, and another opportunity
was lost—an opportunity if not for peace and disarmament,
at least for finding out whether peace and disarmament are
possible.

## CHAPTER VIII

# Conscience in Politics

IT WAS characteristic of Adlai Stevenson's role in the America of the 1950's that in January, 1959, he should have taken time out from partisanship to give the first annual lecture in memory of the eminent liberal minister, A. Powell Davies— and that he should have been asked to do so. Davies, from his Unitarian pulpit in Washington, had for many years effectively argued for deciding political questions in fundamentally moral terms. Politics without moral idealism, he had maintained, is a sham and betrays American civilization. Compromise between just and honest men is necessary; but compromise with evil is intolerable. And Davies had not hesitated to take an active part in liberal causes—working for civil rights and equal opportunity in many groups and movements. For him religion was practical and immediate, the force that introduces moral idealism into behavior.

Stevenson had known and admired Davies. His own convictions were similar, and it seemed appropriate that he should both do honor to Davies and express his own views on morality and public affairs in a memorial lecture in Constitution Hall. Indeed, the occasion permitted him a better opportunity than he had yet had to speak from his heart on the subject which meant most to him and which had led him into politics. "From the mountain of vision," he said,

Dr. Davies constantly proclaimed the political relevance of moral principle and of religion as a "judgment of righteousness." From the dusty plain of politics I would like in my turn to reaffirm this relevance. I like to believe that there may be some value in echoing testimony from a layman who has spent his middle life in the press and confusion of great events in government service, in diplomacy and in politics.

But his testimony was a good deal more than an echo. His text was from Davies:

The world is now too dangerous for anything but the truth, too small for anything but brotherhood.

This had been Stevenson's theme, now in a major, now in a minor key, from the moment of his first appearance on the national scene. In the years he had spent on that scene the reasons for telling the truth, for "talking sense to the American people," had multiplied and were reaching a crisis of compulsive necessity. Stevenson recalled a conversation with Albert Schweitzer in his jungle hospital in Africa a year or so earlier. Schweitzer had told Stevenson that he "considered this the most dangerous period in history, not just modern history, but all human history." Why? Because, Schweitzer had said, "heretofore nature has controlled man, but now man has learned to control elemental forces before he has learned to control himself."

Under these circumstances Stevenson found a fearful contrast between the attitude and spirit he had found on his Russian travels and what he saw and felt at home. The Russians gave him "an overwhelming impression of thrust and purpose in most aspects of life." The Americans seemed content to satisfy their wants at ever higher standards of material

luxury and to long for a peace in which they could be let
alone to pursue their desires. Government in the United
States seemed to be encouraging an atmosphere conducive to
the illusion of security and stability, while the world moved
more rapidly than ever "down the ringing grooves of change."
Stevenson had found the fact of change central everywhere
in the world and consciousness of it characteristic of the
Communists not only in Russia. Serge Obraztsov, director
of the famous Moscow puppet theater, told Stevenson:

> I visited China five years ago. It was the most extraordi-
> nary experience of my life. People in China have had noth-
> ing—nothing! Now several hundred million people are
> dreaming of tomorrow. I cannot describe to you the feeling
> of excitement there—much, much more even than here in
> the Soviet Union.

But the Soviet Union was leading the Communist world, and
its "energy, . . . drive, . . . dedication . . . spill over into
international affairs." The efforts of the Communists, Adlai
Stevenson had seen, "are not confined to areas of Communist
control. They are world-wide, and there is no corner of the
earth's surface which they think too insignificant for their
attention." Trading all kinds of products and raw materials,
providing economic aid, sending out academic men and technical experts all over the world, the Soviet Union was losing
no opportunity to pump out its "flood of propaganda depict-
ing the Soviet millennium of bumper harvests and happy
workers."

But if these things were coming to be common knowledge
in the United States and the West, the critical question was
whether "we try to grasp the scale of dedication that lies
behind" the Soviet effort.

Why should they be so busy? Why so much work and thought? Why such diversion of resources? Why such patience through every setback, such forward thrusts through every point of Western weakness? Heaven knows, we only want to stay home. Why don't they? Why do we never meet an isolationist Communist?

The answer, Stevenson said, was not far to seek. Partly, of course, there was the advantage of trade, and partly the Russians' felt need for friends to provide security.

But the important thing is that the Soviet Russians believe in their truth, as the men of the Western world once believed in theirs. They, not we, are firing the shots that are heard round the world—and also the satellites that orbit above it. The fact that their faith is in many ways an evil perversion of the great propositions that once made the blood course in our western veins does not alter the fact that their tempo is dynamic and ours sluggish—even, I think, to ourselves.

This contrast between the dynamism of the Communist world and the sluggishness of the West, Stevenson suggested, must be caused either by the failure of faith in freedom to justify itself or by the failure of free men to maintain their faith. Communists cannot claim more cogently that their solidarity is the result of faith than can free men. "No country on earth," said Stevenson, "owes the sense of community more explicitly to the fact that it is united not by race or nationality but by fidelity to an idea." And the American idea was never confined in its application to America only. The propositions of "the Jeffersons, the Lincolns, the Woodrow Wilsons—were great because they were able to speak for humanity at large. . . ." The American dream itself could not be blamed for the slackness of the Americans. "Its truths," Stevenson continued, "are still 'self-evident.' The possession of liberty and the pursuit of happiness—rightly understood —have not been overthrown as the highest goods of human society." These ideas were still at work, and to the danger of the Communists, in many places throughout the world. Poland was a good example that Stevenson had recently seen for himself.

But if the ancient liberal faith was not at fault, why had its adherents, Stevenson wondered, failed it? "Why this lack of initiative? Why this paralysis of will?" Why had American action been so depressingly "defensive"?

> We have offered aid not to help others but to shield ourselves. We have reacted to countless Soviet initiatives; acted on our own initiative barely at all. We watch the skies for other people's sputniks and listen to the telegraph wires for other people's moves. Yet we are the free men of this universe, the children of liberty, the beneficiaries of unequalled abundance, and heirs of the highest, proudest political tradition ever known to man!

Stevenson now turned to an effective figure to push home the problem he was addressing. Himself for years the "conscience" not only of his party but of his country, he now called for "an examination of what you might call our collective conscience" as a more important exercise at that juncture of history than any other:

> You can have a perfect assembly of pieces for your watch, but they are worthless if the mainspring is broken. I am not basically worried about our various pieces—our technology, our science, our machines, our resources. But I am concerned, desperately concerned, about our main-

spring. That it has run down, we know. But is it broken beyond repair?

The answer was that the break was not irreparable. But the watch had certainly run down. "In recent years," Stevenson said, "we were stifled with complacent self-confidence. We believed ourselves dominant in every field. We talked of the 'American Century.' We forgot the ardors and efforts that had given us a measure of pre-eminence." The great trouble, Stevenson continued, was our tendency to confuse "the free with the free and easy." But the condition of freedom is not only rare today, it has always been rare:

> If freedom had been the happy, simple, relaxed state of ordinary humanity, man would have everywhere been free —whereas through most of time and space he has been in chains. Do not let us make any mistake about this. The natural government of man is servitude. Tyranny is the normal pattern of government. It is only by intense thought, by great effort, by burning idealism and unlimited sacrifice that freedom has prevailed as a system of government. And the efforts which were first necessary to create it are fully as necessary to sustain it in our own day.
>
> He who offers this thing we call freedom as the soft option is a deceiver or himself deceived. He who sells it cheap or offers it as the by-product of this or that economic system is a knave or a fool. For freedom demands infinitely more care and devotion than any other political system. It puts consent and personal initiative in the place of command and obedience. By relying upon the devotion and initiative of ordinary citizens, it gives up the harsh but effective disciplines that underpin all the tyrannies which over the millennia have stunted the full stature of men.

The "discipline of democracy," as T. V. Smith has called it, has always been difficult to maintain. Stevenson had good reason to know this from his own experience as a candidate for the Presidency and as leader of a political party. He had felt the intensity of the "battle for the candidate's mind" in which he is battered and cajoled to take positions which will flatter the people and so, hopefully, produce votes. Promises of ease and relaxation from the tensions of the world had characterized the two Republican campaigns which had defeated him, and he had been forced to struggle against forces in his own party urging him to "hit the people" with the "gut issues" of bread and circuses. But from the moment when he had asserted that "self-criticism is the secret weapon of democracy" and had promised to "talk sense," Stevenson had resisted such forces. He had, in fact, become known precisely for his resistance, and his place of esteem and leadership was owing to his ability to identify his own with the public conscience. In his Davies lecture he compressed the whole persistent meaning of conscience in politics into a single paragraph:

I believe we have had enough of adjustment, conformity, easy options and the least common denominator in our system. We need instead to see the "pursuit of happiness" in terms which are historically proven and psychologically correct. The dreary failure in history of all classes committed to pleasure and profit alone, the vacuity and misery accompanying the sole pursuit of ease—the collapse of the French aristocracy, the corruption of imperial Rome, the decline and fall of the resplendent Manchus—all these facts of history do not lose their point because the pleasures of today are mass pleasures and no longer the enjoyments of an elite. If we become a nation of Bourbons,

numbers won't save us. We shall go their way. Vacuity
and indifference are not redeemed by the fact that everyone can share in them. They merely restrict the circle
from which regeneration can come.

As he pursued this theme Stevenson's eloquence rose as it
had so often during his early years on the national scene—
and less often as his role had forced him more and more into
the conventions of public speaking. He now had hold of a
theme which could open the way for full expression of a
moral indignation rare in national politics but indisputably
the measure of his own public meaning. "We do not slip into
happiness," he said, "it is strenuously sought and earned.
A nation glued to the television screen is not simply at a
loss before the iron pioneers of the new collective society.
It isn't even having a good time." The inordinate consump-
tion of "drink and tranquilizers" reminded him of a remark
made about the court of Louis XIV by the French moralist
La Bruyere: "Its joys are visible, but artificial, and its sorrows
hidden, but real."

The trouble, Stevenson thought, was a "misunderstanding
of the real nature of freedom." As he proceeded to correct
that misunderstanding, according to his own philosophy, he
placed himself in the center of the humanistic tradition. Perhaps no political leader of this generation, unless it be Nehru
of India, has so forcefully asserted the necessity of inner
control over the passions as the indispensable contribution
of the citizens in a free society. In his autobiography, *Toward
Freedom,* Nehru had many years before written:

It has been, and is, man's destiny to control the elements,
to ride the thunderbolt, to bring the raging fire and the
rushing and tumbling waters to his use, but most difficult

of all for him has been to restrain and hold in check the passions that consume him. So long as he will not master them, he cannot enter fully into his human heritage.*

Stevenson made the same point in a different, less flamboyant key. But a common insight and a common sense for the imperatives of freedom in the crises of the twentieth century brought the Eastern and the Western statesmen into spiritual accord:

> How are we to defend freedom if, for the tyranny of external control we substitute the clattering, cluttering tyranny of internal aimlessness and fuss? This freedom for our souls, freedom at the profoundest level of our being, is not a gift to us by our contemporary way of life. On the contrary, much of this life is a direct conspiracy against it. And if we cannot—by a certain discipline, by readiness for reflection and quiet, by determination to do the difficult and aim at a lasting good—rediscover the real purpose and direction of our existence, we shall not be free. Our society will not be free. And between a chaotic, selfish, indifferent, commercial society and the iron discipline of the communist world, I would not like to predict the outcome. Outer tyranny with purpose may well triumph over the inner, purposeless tyranny of a confused and aimless way of life.

Stevenson now offered three benchmarks to test whether the Americans would be capable in this century of regaining the sense of national purpose and the discipline to carry it out. The first was what he called "remediable poverty." The "affluent society" is normal today for a great majority of

---

* Jawaharlal Nehru, *Toward Freedom,* New York: The John Day Co., 1941, page 263.

Americans to a greater or lesser degree. But some five million families do not share in it at all. This poverty, Stevenson argued, must be wiped out, and it will be wiped out "only if the well-to-do majority of today do not repeat the selfish indifference which, in many communities, has been the epitaph of yesterday's wealthy elite."

Second in the list was "the rights and status of our colored citizens." Stevenson pointed out that the "four hundred years' dominance of men of white skin is ending." The problem of discrimination in America is part of a world problem, but the American share of it is susceptible of remedy more readily than elsewhere. If America is to save herself, Stevenson maintained, the civil rights of all citizens must soon be protected not only by law but by custom. And this can never be accomplished "unless there are enough white men and women who resist in the core of their being the moral evil of treating any of God's children as essentially inferior."

Finally, Stevenson pointed to the fact that 16 per cent of the world's peoples, the Atlantic world, consume 70 per cent of the world's wealth. The United States, for her part, consumes the lion's share of that. "To the moral implications of this gap," Stevenson went on, "we cannot be indifferent." It will require "moral insights of justice and compassion" to stir us into an understanding of the privileged position which sets us apart from the rest of the world. For, said Stevenson, "We are not going to be stirred by our own needs."

Such considerations raise inescapable questions about the relevance of moral virtue to politics. What is practical? Josiah Royce had once argued that "only the eternal is practical," and Emerson had still earlier warned his fellow Americans that only ideal ends are worth serving. But the Emersonian gospel was long out of fashion as Stevenson spoke, and Royce had been forgotten. Stevenson was nevertheless prepared to face the logic of his criticism. "You may argue," he said, "that these qualities—of dedication and selflessness—are pretty remote from the realities of politics." Again, "Ambition, drive, material interests, political skills, the art of maneuver—all these, you say, have their part, but do not let us pretend that the democratic process is primarily a school of virtue or an arena of moral combat." His answer to such rhetorical questions was in effect to say that democracy had indeed better be an "arena of moral combat," and to suggest that perhaps, after all, it is.

And yet, I wonder. It has been the view of great philosophers and great statesmen that our system of free government depends in the first instance upon the virtue of its citizens. Montesquieu made virtue the condition of republican government; Washington declared that it could not survive without it. We have had a hundred and seventy-five years of it since their time and no one can deny that the system has survived a remarkable amount of skull-duggery. In fact, it is probably a tougher system than its founders imagined. Yet I believe they are right. For no democratic system can survive without at least a large and active leaven of citizens in whom dedication and selflessness are not confined to private life but are the fundamental principles of their activity in the public sphere.

But if it be agreed that free government does indeed require substantial numbers of citizens gifted with these virtues, the problem remains how to persuade them to come forward. There is never any difficulty in getting *interested* persons to be active. "We do not need societies for the promotion of lobbies," Stevenson pointed out. "Nor in any generation," he continued, "do we lack politicians whose only principle is the advancement of their own career—the starry-eyed opportunists and all the other eager men in a hurry to the top."

The point is to get *disinterested* citizens to be active as leaders and as followers. Partly, the way must be simply by exhortation, by appeal to conscience. Partly, however, there is an appeal to a kind of interest—the public interest itself. People do want reform, and they can be and are persuaded to work for it:

But there has never been any disinterested reform without disinterested reformers. And here we come to the essential contribution made by dedication and selflessness to the public good. No one ever did any good in politics without readiness for endless hard work—for the grinding, boring, tedious work, as well as the glamorous, high sounding, headline hitting work. The painstaking hours collecting the facts, the hours in committee and conference, the hours in persuasion and argument, the hours of defeat and disappointment, the hours of disgust and revulsion at the darker sides of human behavior—these cannot be supported without energy and devotion. No reforms come easy; even the most obvious will have its entrenched enemies. Each one is carried to us on the bent and weary backs of patient, dedicated men and women.

In a world pressing so relentlessly to flatter the personal desires and whims of the people, in a society so devoted to cultivating "consumer wants," Stevenson observed that "it takes an extra dimension of vision to see beyond our inner circle of interest."

He could not and did not conclude on an optimistic note. He recalled the youth of such men as Lincoln and Douglas, and the attention and concern they drew in their manhood when they debated the question of a nation "half slave, half free." He could not believe that Americans today are more seriously engaged, more necessarily involved in their own pursuits, so that the moral slackness of the twentieth century is somehow forgivable:

> In a century in which so many mentors of the public mind —from the psychiatrists to the ad-men—speak to us in terms of "what we owe to ourselves," may there not indeed have been a slackening of devotion compared with those days, not so long distant, when what one owes to God and his neighbor was a common theme of public discourse?

The lecture was not, perhaps, a masterpiece. The writing was somewhat uneven and the substance occasionally repetitive. To many it would seem hopelessly didactic and tedious. To others it would suggest once again the utopian dreaming of the "egghead." But it contained many passages of great power and eloquence, and its sincerity and immediacy were compelling to any who were willing to look with the speaker beyond their own interested preoccupations. It is hard to imagine any other leading politician of the twentieth century making such an address. Even Wilson, who had the learning and the understanding to do so, lacked the humility which was its first prerequisite. This address, more than any other of his many speeches and writings, explains why Adlai Stevenson could not have been elected to the Presidency of the United States in the "Age of Eisenhower." But it explains, too, why he had established for himself a high place in the history of his time and his country, and why, on transcendent issues, he could and did articulate the will and the vision of America as no other of his contemporaries could do.

The Davies lecture was widely reprinted and drew a good deal of editorial comment. It was the recipient of lavish praise from newspapers and magazines which had never missed an opportunity to attack Stevenson when he was a candidate for office. The press, which had so long guarded Eisenhower like an impregnable fortress, seemed quite willing to agree with Stevenson that American civilization was in a state of moral decline. But it seemed somehow impossible for the editors to connect the Eisenhower administration with such a decline. No one observed that Stevenson had warned of it for years, while Eisenhower presided over it and symbolized its spirit of complacent self-satisfaction. *Life,* for example, headed a full page editorial "The Cost of Easy Options" and called Stevenson's address "the best recent statement of this informed worry"—about the moral condition of the country and its fitness to survive against "Russia's iron purpose." The editorial quoted many passages from the lecture with approval and associated *Life* with each of Stevenson's main criticisms of American life and with his conclusions. In their words, "The prevailing climate of moral isolationism is found in every corner of our society, and almost everywhere it contrasts sadly with the immoral but energetic expansionism of the Soviet Union." Like so many others, *Life's* editors could recognize in Stevenson the authentic voice of conscience in politics, except at election time.

## CHAPTER IX

# 1960: End of an Era

I

IF 1960 was to mark the end of Adlai Stevenson's leadership of the Democratic party, it was nevertheless another indication of his extraordinary hold upon the minds of many people that few prophets cared to eliminate him from contention for the presidential nomination until the very moment of John Kennedy's ascendance in July.

Stevenson was never a candidate. Shortly after the 1956 election he had stated emphatically that he would not again seek his party's nomination, and thereafter he made it clear many times that he thought it best that he be succeeded in 1960 by a younger man with a "freshness and eagerness for the fray" which he himself no longer felt. Long before the season of primaries began he found himself under heavy pressure to name and support a favorite. Hubert Humphrey of Minnesota and Kennedy had, perhaps, particular claims to his support, since they had closely associated their views with his and had been vigorous leaders in his cause in earlier years. But partly, at least, because both men were well qualified and partly, also, because he did not wish to intrude or to adopt a divisive role in the party, Stevenson refused to make a choice. He spoke highly of all the candidates when occasion called for his comment—including Senators Stuart Symington of Missouri and Lyndon Johnson of Texas.

But inevitably his position of detachment from the competition, underscored by his departure early in the year for a long tour of Latin America, suggested his own availability. And his partisans in nearly every state went to work to "draft" him at the convention if circumstances should arise in which no avowed candidate could command a majority of the delegates. As early as January 8 the *New York Times* reported that "a national movement to draft Adlai E. Stevenson for the Democratic nomination for President is being sponsored by his Midwest supporters." The *Times* went on to report the launching of an Ohio committee and a move to put Stevenson's name in the Oregon primary. At the same time, the *Times* carefully included disclaimers by Stevenson himself and his close associates that he would enter any primary or assist any draft movement. Stevenson, indeed, gave no help to these supporters, then or later. Some of them were close personal friends and associates, and some were party leaders who had stood behind him in the past. But he gave them no encouragement. There is reason to believe that at the moment in May when the Paris summit conference collapsed and the world was bewildered by the U-2 incident, a clear signal from Stevenson might have brought an onrush of support sufficient to win the nomination. But no signal came—at least no unambiguous signal.

For Stevenson's behavior throughout the first six months of 1960 always appeared ambiguous. In many ways he acted like a candidate. He was known to question the qualifications of Richard Nixon, the probable Republican nominee, and to be confident that he could defeat him. It seemed likely that a man who had twice campaigned so intensively for the Presidency would still cherish the hope to attain it, and wherever he went abroad or at home he heard himself described as the man best equipped to serve. His availability, as a dedicated public servant, was never in doubt. If nominated he would run, and if elected he would serve. But he would not be a candidate.

In the end, of course, he became a candidate in spite of himself. As he put it on a national television program on the eve of the convention, referring to the people of the draft movement, "I am *their* candidate." By then, perhaps, it was too late. But the triumph of Kennedy at Los Angeles was no defeat for Stevenson, for the convention, regardless of its outcome, was certainly what he himself called his "finest hour." Its spirit, its platform, its deliberations, and all its speeches were Stevensonian. In eight years he had put as clear a mark upon his party for his time as had Franklin Roosevelt or Woodrow Wilson for theirs. And never was he more effective than in the six months just preceding the 1960 convention.

II

Before his departure for Latin America, Stevenson, very much in the manner of a candidate though formally in his character as party leader, published in *Foreign Affairs* a program for action which he called "Putting First Things First." Later, this article, together with other recent addresses, was published as a book under the same title—Stevenson's sixth book in seven years. The article served as a kind of platform on which he could stand in the months before the party would choose its new leader—or retain the old. In any event it could and did serve the party as a pre-convention manifesto to help draw the lines at issue with the Eisenhower administration and the future Republican leadership.

Stevenson's program was advanced as means to peace—
"the most imperative business in the world today." Under
such headings as "Economic Development," "The Atlantic
Community," "Arms Control," "China," "Europe and the
Middle East," and "A Sense of Purpose," he analyzed the
existing situation and suggested action for the future. He
put economic development first because, as he had so often
pointed out, it is the gap between an average annual income
in the United States of more than two thousand dollars and
an average income among one-third of the world's people
of less than one hundred dollars, which underlies most, if not
all, phases of the twentieth-century crisis. To move vigorously
and purposefully and wisely toward closing that gap would be
the first great step toward peace. He noted five conditions
for a successful program:

> We shall be engaged on this program for at least 40 years.
> We shall require a professional staff, with the languages
> and skills needed in this whole new field of activity. In-
> formed opinion tells us that at least $5 billion a year is
> needed—from all sources, public and private, domestic
> and foreign. We shall have to coordinate all aspects of the
> effort with other nations—not only investment but oppor-
> tunities for trade, international liquidity and so forth. To
> get the maximum results the developed nations must all
> cooperate. The time has certainly come for other coun-
> tries to share more of the common burden of assistance.
> In such circumstances the United States cannot expect to
> have full control of the use of all its expenditures for
> development purposes.

Stevenson was not optimistic about the progress of such a
program under conditions then prevailing. In particular, he
thought, there was "still more than a hint that if the Commu-

nists would behave, the economic development program could be cancelled."

Since his own participation in planning for NATO meetings in 1957, Stevenson had never believed that the administration's attitude toward the Atlantic community was either sufficiently positive or politically realistic. "Beneath the surface," he now wrote, "there are dangerous cross-purposes within our alliance. They are not to be overcome by ceremonial travels and by hasty diplomacy in preparation for meetings with Mr. Khrushchev." In the economic field he pointed out the gross contradiction between our "Buy American" policy and the liberal, free-trade principles we were urging upon our allies. In the military field he expressed concern at the diminishing of NATO's deterrent capacity, and observed that disarmament can only be talked seriously when there is "equality of strength and equality of risk." In addition, there were costly duplications to be removed and conventional forces to be built up on a basis of proper proportion among the allies. In sum, he advocated the establishment of "an Atlantic Council with real powers" to formulate joint policies "for sharing our responsibilities and bringing about a genuine and equal partnership between the United States and Western Europe." Before a summit conference was held with the Russians, the West should make its own plans and move forward with its own programs:

> I believe a North Atlantic Conference should be held to outline new common policies for defense, disarmament, space exploration, monetary reserves, tariffs and a larger economic sphere, and aid to the underdeveloped areas, giving, I hope, new terms of reference to NATO and to other organizations. I think Europe should take the initiative toward creating some such new organization to deal

with our great and growing problems and to promote more systematic Western cooperation.

Turning to the problem of arms control, Stevenson set forth his view that the "root of East–West tension is fear." It is really inconsequential whether fear is justified by the behavior of one party or the other; what matters is that fear is a fact. For this reason he felt that to insist that political settlements be reached before arms reduction takes place is to guarantee that there will be no arms reduction. Only by the limitation and control of armaments, he thought, could a climate of international feeling be achieved in which some political questions might be successfully addressed. His own experience of Russian leaders and his observations of Russian attitudes led him to believe that the Soviet people and "some, at least, of the Russian leaders" were sincere in their desire for disarmament, at least to the extent of halting the testing and development of nuclear weapons. This, as he had so many times argued, would at least be a start. Calling attention to the United Nations disarmament resolution which recognized that disarmament would promote trust among nations, he summarized his own view in these words:

In short, it looks as though controlled disarmament was back at the top of the world's agenda where it belongs. I am sorry that the United States did not take and hold the lead as I urged in the 1956 presidential campaign. The recent proposal by some of our leaders that the United States resume underground nuclear tests, just when the first break in the arms deadlock seems possible, shocked me. I can think of few better ways to chill the prospects, deface our peaceful image and underscore the Communist propaganda that they are the peacemakers and we the warmongers. We should extend our test suspension so long as negotiations continue in good faith and Russia maintains a similar suspension. The good faith of the negotiations is decisive, because indefinite suspension amounts to a test ban without inspection.

There had always seemed to be a certain ambiguity in Stevenson's attitude toward inspection of nuclear test suspension. At first, early in 1956, he had pleaded for suspension of H-bomb tests without inspection, on the proposition that such explosions could immediately be detected anywhere in the world. Later, perhaps because his views were seriously distorted by both President Eisenhower and Vice President Nixon, he seemed at times to take the view that inspection agreements must be reached before suspension could be safely undertaken. In 1958 and 1959, however, he criticized the administration for putting its premium on inspection systems to which the Soviets would not agree, thus making test-suspension agreements impossible.

Stevenson's inconsistency on this issue was more apparent than real, however. Putting together his various statements on the whole matter, from 1956 to 1960, provides a clear series of related propositions which would make for a consistent American policy:

(1) The United States should suspend its tests of H-bombs and challenge the Soviet Union (and Great Britain) to do likewise. This should be done to rid the world of the fear of poisonous fallout.

(2) Cessation of H-bomb tests would have the effect of breaking the deadlock in the arms race. It would be a start.

(3) The military risk would be minimal since H-bomb explosions can be detected immediately anywhere in the

world, and the United States could immediately commence a new program of tests if another power began testing.

(4) The United States should take the lead in attempting to secure a test suspension agreement for all nuclear weapons. Such an agreement ought to include measures of international inspection, but inspection should not be the aim of the negotiation, since the Russians would always object.

(5) Voluntary suspension by all parties during the period of negotiation should be the actual means of achieving agreement—since prolonged negotiation accompanied by prolonged suspension would buy time and habituate all parties to mutual confidence.

(6) Agreement to nuclear suspension with international inspection should lead to agreements banning the development and manufacture of nuclear weapons and to a system of inspection and control of all nuclear production.

What distinguished Stevenson's policy was thus precisely the fact that it *was* consistent—that is, it aimed always at the relief of tension caused by the fear of nuclear war—while both the Russian and the official American positions were frequently altered for the evident primary purpose of maintaining political and military advantage. What progress there actually was in the 1950's seemed to arise from response to popular pressures of the sort Stevenson himself continued to articulate.

In his *Foreign Affairs* article Stevenson next turned to the thorny question of China. Approaching the role of China in terms of the disarmament problem, he suggested that the Soviet Union might have a compelling interest in moderating the attitude of Chinese leaders to avoid the risks of a world war touched off by Chinese aggressiveness. In particular, he proposed support of Khrushchev's proposal for an "atom-free zone" in the Far East. Stevenson's suggested terms for a settlement in the Far East he summarized as follows:

> On the Communist side, the concessions would include the extension to China of any system of international inspection of disarmament, ending the threat of force against Formosa and subversion in Indo-China, a peaceful frontier settlement with India, free elections under United Nations supervision in Korea, and acceptance of the right of the inhabitants of Formosa to determine their own destiny by plebiscite supervised by the United Nations. On our side, concessions would presumably include an end to the American embargo on China's admission to the United Nations (not to be confused with diplomatic recognition), the evacuation of Quemoy and Matsu and the inclusion of Korea and Japan in the atom-free zone and area of controlled disarmament.

Stevenson went on to state unequivocally the unanswerable argument, so distasteful to many American politicians and conservative columnists, for China's admission to the United Nations:

> . . . It is clear that no general control of disarmament has any value unless it includes China, and it is difficult to see how China can accept international control when it is not, formally, a member of international society.

In addition, Stevenson thought that China would be more accountable to world opinion as a member of the United Nations than "as an outcast." This view, answering as it did to the imperatives of the actual world situation, was not calculated to rally party politicians to his support for a third nomination, and may serve as a measure of Stevenson's sincerity both as a non-candidate and as a student of world affairs.

Stevenson touched only lightly on the Middle East in this 1960 manifesto, asking again for efforts to embargo arms shipments, suggesting that the possibility of an atom-free zone in the Middle East be explored with the Soviet Union, and that the United States call upon the Soviet Union and the whole United Nations to try to harmonize relations between Israel and the Arab states.

For the immediate future Stevenson thought that Europe, especially Germany, would continue to provide the "critical point of tension." Again he looked to disarmament as the only real hope for European settlement. He estimated that the Russians would run the greater risk in the event of withdrawal by both East and West, and under such circumstances saw little hope for political agreements other than "postponements of the problems of divided Europe."

Finally, in a clear echo of his Davies lecture, Stevenson argued that our tendency to "hasty improvisations and snap decisions" could only be stopped and replaced by firm purpose in world affairs if we treat our problems at home with firm purpose:

> The truth is that nations cannot demonstrate a sense of purpose abroad when they have lost it at home. There is an intimate connection between the temper of our domestic leadership and the effectiveness of American influence in the world at large. President Wilson gave a profound new direction to international thinking because he was a pioneer of the New Freedom at home. President Roosevelt's universal prestige as a liberal force in the world was deeply rooted in the New Deal, and this was the tradition carried on by President Truman in such great

ventures as the Marshall Plan and the Point Four program. The link is no less vital today. If we cannot recover an aspiring, forward-looking, creative attitude to the problems of our own community, there is little hope of our recovering a dynamic leadership in the world at large. By our default as much as by his design, Mr. Khrushchev is enabled to continue dictating the terms of the world's dialogue.

Thus, as perhaps he intended, Stevenson struck in January the keynote of the 1960 Democratic presidential campaign. Presently he himself departed for South America on a long and exhausting study tour of the sort he had made in prior years to nearly all other parts of the world. But at home Hubert Humphrey and John Kennedy, contesting the Democratic presidential primaries, conducted their campaigns in terms of the Stevensonian intellectual substance, if not the Stevensonian manner. And after the convention in July, John Kennedy carried the same plea to the people in his successful campaign for election.

## III

On the eve of his departure for Latin America, February 9, Stevenson issued a statement to the press describing his journey as that of a "learner." Pointing to such growing problems of Latin America as "population rise, inflation, the shortage of capital and confidence, economic and political instability and fluctuating export prices," Stevenson asserted the need to strengthen our political, economic, and cultural ties with Latin America as means to advance the "deep commitment" of all Americans to democracy and freedom. He hoped, he said, to return from his trip "a much better citizen of the hemisphere." If his accent on citizenship was unusual for a statesman about to travel abroad (even a statesman out of office!) it was nevertheless characteristic of Stevenson, and characteristic of his foreign tours throughout the 1950's. Comparisons are not, perhaps, helpful in such matters, but it is reasonable to doubt whether any American party leaders have made greater efforts to convert their travels into the kind of knowledge that can in turn be translated into wise policy.

Stevenson's Latin-American tour followed the same pattern of organization he had used previously but with more advance planning. Before his departure Stevenson not only had the advantage of briefing sessions with State Department specialists on Latin America, but also sat in lengthy conferences with a group of university scholars who discussed long-range Latin-American problems with him. He took with him Carleton Sprague Smith of New York University, a distinguished student of Latin America, William Benton, an old friend who as Assistant Secretary of State had founded the Voice of America and had built a solid reputation as a student of American foreign policy, and William McC. Blair, Jr., his companion and assistant on many tours.

In each of the capitals Stevenson was received by the head of government and shown the most cordial friendship, not only by officials but also by crowds of plain people who came to see him and to display their sense that he was the authentic representative of that American "Good Neighbor" policy identified in Latin America with the Democratic party since the time of Franklin Roosevelt.

In sharp contrast with the Nixon "good-will" fiasco of two years before, Stevenson was on several occasions accorded popular receptions which, on the testimony of the Latin Americans themselves, set new records for size and friendliness. In Bogotá, for example, people moved into the city from as much as fifty miles away, filling the roads with traffic for two days, to see Stevenson and hear him say a few words in the public square. In a report appearing in his nationally syndicated column, Ralph McGill described the enthusiastic reception as follows:

> Stevenson arrived at Bogotá. He and his group were whisked, so to speak, from airport to bull ring. A massed group outside, which was in a near state of rebellion because it could not obtain tickets, set up a great cry of acclaim. Inside, the massed benches let loose such a roar on Stevenson's entry he flinched, thinking nothing less than a riot had erupted. His admiring hosts, the president and officials, assured him it was his welcome.
>
> Speeches were made. The great gates of the arena opened for the ritual of the bullfight processional. It brought further salutes to the visiting American.
>
> Nor was this the end. The matadors dedicated their bulls to him. And when the long hours of death in the afternoon were done, Stevenson was hoisted to the shoulders of the crowd, along with the three matadors, and carried about the ring to the vast delight of the multitude which kept up a Niagara of shouts. Chief among them was "uno," meaning the first or the best.
>
> Here was a picture Americans would like to have seen in their newspapers and on their television screens. Stevenson described himself as somewhat nervous. Being ridden on the shoulders of excited admirers is even more insecure and precarious than a perch on a camel. But he liked it. It was good to be a national hero instead of coping with one. At that moment the life of a top-flight matador seemed to him much more enjoyable and happy than that of a presidential candidate.

But beneath the surface of ceremony and personal popularity, hard work was going on. Stevenson meant to learn, and devoted every moment he could find to observation and to conversations about the tough realities of Latin-American life with people of as varied experience and condition as time and circumstance permitted. Smith, Benton, and Blair contributed their experience to a common pool for Stevenson's use.

Stevenson returned from Latin America with a new sense of urgency. The difference between the facts of Latin-American life and the understanding of them in the United States, and the difference between the realities of United States relations with Latin-American nations and the administration's picture of them, seemed to Stevenson tragic differences. As he put it in an article for *Look,* published after the election campaign was over:

> I was in Latin America last spring at the same time as President Eisenhower. I traveled through twelve countries in eight weeks. The President went to four countries in ten days. He came back optimistic. I came back deeply concerned.

He continued, listing five sources of concern:

> I am concerned because Latin America is in social and political revolution—like most of the evolving regions of the world.
>
> I am concerned because, in a region rich in resources, half the people are hungry, half don't sleep in beds, half are illiterate.
>
> I am concerned because the population increase is the fastest in the world and is outstripping production.
>
> I am concerned because of our ignorance about our Latin neighbors and because of the anti-Americanism I found, in spite of their moving welcome to me.
>
> And I am concerned that if they don't achieve their

desire for a better economic and political life, we may
find enemies, not friends, on our doorstep.

As for the "Good Neighbor" policy, Stevenson found a bitter
joke going the rounds: "We are the good; you are the
neighbor."

The compelling need, Stevenson concluded, was to reverse
the trend toward anti-Americanism. The first step would be
to understand clearly that "it is foolish for us to attribute
anti-Americanism just to Communist agitation." The anti-
American criticisms he met with most consistently were not
Communist-inspired but inspired by the facts of inter-Ameri-
can relations. He summarized these impressions in a list of
five recurrent Latin-American criticisms:

(1) That the Eisenhower administration "has been
basically concerned with making Latin America safe for
American business, not for democracy," and has supported
"hated dictators."

(2) American businessmen are "interested only in the
profits they can make, not in the country and its development. . . . They are patronizing. . . ."

(3) The United States has "neglected Latin America
since the war." Money has been given in large sums to
"neutrals and even enemies" for development but the
Americans only make loans to Latin America "due to
pressure or fear of communism."

(4) The United States is "blamed for the low prices
and the price fluctuations of copper, coffee and other products . . ." leading to cycles of boom and bust.

(5) There is a "psychological irritation." It is easier
to blame the United States for troubles than to accept the
responsibility, "and the Communists channel the resent-

ments and frustrations into hostility against a big, rich, obvious target."

Such a list, Stevenson thought, must be taken seriously. If the criticisms are exaggerated in most cases, there is nevertheless an important element of truth in each of them. It must be a matter of the deepest concern that a people so devoted to freedom and democracy tend more and more to look upon the Soviet Union, or even China, as the best example of economic development. The United States has the power to reverse such tendencies:

> Once the United States and its sister republics begin to cooperate in earnest, I have no doubt our relations will quickly improve. The problems themselves can be tackled only by the Latin Americans. *They* will have to take the bold, brave, difficult steps to achieve better land use and distribution, better housing and better education at all levels. *They* will have to clear away the ghastly slums that surround every city. *They* will have to reform taxation and tax collection, cut corruption, reduce the waste on arms, increase the rate of savings and narrow the gap between rich and poor.
>
> In short, Latin America will have to make its own New Deal. But we, in the meantime, can do much to help.
>
> We can loudly and clearly declare that we believe free and democratic societies are capable of providing rising standards of living and that we are prepared to prove it in cooperation with those who will help themselves.
>
> We can propose, not a "Marshall Plan" for Latin America, but a *marshaling* plan—to marshal the available resources of the hemisphere for a carefully planned and sustained attack on economic stagnation and those old

scourges—illiteracy, poverty, and hunger—that we have all but banished from North America and that must now be banished from South America.

I repeat: *They* must do it. But we can help, first with surveys and the essential planning on a national and regional scale; second, with financing on a long-term basis of sound projects broad enough to assure progress; and third, with technical skills, experts and teachers in everything from agriculture to unionization.

In his own behavior during his tour Stevenson gave an indication of the kind of manner and attitude he believed could win the respect and co-operation of Latin-Americans. His address at Bogotá, for example, where he received an honorary degree, stressed his theme of hemispheric citizenship. He took pains to show his appreciation of the old Colombian tradition of close kinship between the humanities and government, and to recognize the devotion of Latin Americans to individual liberty and democracy; he displayed an understanding of the rapid political changes toward democracy and away from dictatorship which have marked the past decade, and his sense that "our solidarity and common purpose to defend these [democratic] convictions is the mightiest weapon of the free world."

It is revealing of Stevenson's state of mind as he reflected on his experiences and thought about what he could sensibly say to his Latin-American audiences that he chose to echo his Davies lecture—he quoted Davies ("the world is now too dangerous for anything but the truth, too small for anything but brotherhood"), and cited again the view of Schweitzer that this is "the most dangerous period in history." Finally, he appealed to the Latin Americans "to take the lead in arms limitations." He suggested that Latin America could, in fact, set an example for the world. "I would like to see," he said, "all the Latin-American republics declare as one their intent and determination to avoid an arms race, and to progressively reduce the arms burden in this continent."

## IV

Stevenson's Latin-American tour was less well covered by the American press than any of his previous journeys, not because it carried less public interest but because, as he discovered, Latin America is so little covered. During one stretch of almost a month he never ran into an American reporter. It must have been something of a surprise to him, therefore, when he returned to New York on April 11, to find himself once more the center of attention among newspapermen.

During Stevenson's absence the Democratic nomination contest had been chiefly fought out by Senators Hubert Humphrey and John Kennedy, but with growing indications that Senators Stuart Symington and Lyndon Johnson would presently become candidates before state conventions at least, if not in the primaries. On April 5 Kennedy won his first important victory, defeating Humphrey in Wisconsin. But it was by no means a decisive showing, and many experts were persuaded that neither Kennedy nor any of the other active candidates would be able to get a majority at the convention in July. Stevenson's friends, while respecting his wishes not to become a candidate, were nevertheless anxious to keep him as much in the news as possible, confident that he would accept a draft if one should develop out of a deadlock. While Stevenson's return from Latin America would have been newsworthy in any case, his supporters in New York and Washington made every effort to make sure that it was well-covered and that his appearance before the country was as partisan as the coming of a campaign would allow.

At his news conference on April 11, when asked directly whether he would be a candidate, Stevenson gave a categorical reply in the negative. But on questions about the possibility of a draft he adopted the position that he would not discuss it at all, since if he said he would accept a draft he would be understood to be courting one, while if he said he would refuse a draft he would be called a "draft evader." And on this position he stood firm until the convention.

The effort to nominate the non-candidate continued in spite of Stevenson's reluctance. This remarkable popular movement has no parallel in American history. The only twice-defeated candidates ever to receive a third nomination were Henry Clay and William Jennings Bryan, both of whom worked mightily to get it. No others were even considered for a third try. After his second defeat in 1948, for example, the only further mention of Thomas E. Dewey in presidential politics was as a supporter of Eisenhower or as a possible member of a Republican cabinet. Yet immediately after the votes were counted in 1956, Stevenson supporters were saying that since no one could have defeated Eisenhower this second defeat should not be held against Stevenson, that he could beat Nixon, and that he should run again. The trial runs of the Gallup Poll throughout the second Eisenhower administration showed Stevenson consistently ahead of other possible Democrats and usually ahead of Nixon, the prospective Republican candidate. While Stevenson's strength seemed to dwindle somewhat under the impact of the Kennedy and Humphrey campaigns in late 1959 and early 1960, as late as March he was still running close to Kennedy, and an independent survey by Louis Bean showed that on the basis of switches by 1956 Eisenhower voters to Stevenson in 1960, Stevenson would defeat Nixon by a large margin. The Wisconsin primary, just over as Stevenson returned from Latin America, left matters still in doubt.

The force of Stevenson's presence in American presidential politics was quickly felt. At Charlottesville on April 12 he gave the Founder's Day address honoring Jefferson but lost no opportunity to use Jefferson as a club to beat the Republicans. He asserted, for example, that Jefferson

> would see that our national leadership has not prepared us for the tasks of this searching century; that it has not summoned us to our duty; that it has not, in his words, "kept alive our attention." Too often—and I wish I could call Jefferson as a witness—our leadership has been hesitant and half-hearted, and has concealed from us the nature and dimensions of the crisis.

He went on to deal with "concealment" as a theme. He reviewed the attitudes and behavior of the Eisenhower administration from the "unleashing" of Chiang Kai-shek down to the President's anger "when some of our distinguished citizens and generals express concern about the obvious fact that our defenses are not as strong as they were." He used strong language—the language of a committed candidate, it seemed to many people:

> But these impostures also derive from misunderstanding or disrespect for our system—from a vague feeling that the best kind of government is one in which the people turn their hopes and fears over to a kind of caretaker for the national welfare and conscience, to a benign chief magistrate who countenances little criticism and comforts the people with good news or none.

Again, contrasting the moment of Jefferson's triumph in 1800 (when "there was no lethargy . . . no confusion about our values or objectives") with the present, Stevenson struck hard:

Today we are no longer poor and defenseless. We are by far the richest nation on earth and, until recently, the most impregnable. Yet, ironically, our actions have been timid and irresolute. Our leaders talk of freedom—and embrace dictators. We do not act as frightened as we did during the shameful McCarthy era. But to millions of people just emerging from feudalism or colonialism we still look like a nation that has forgotten its revolutionary heritage and moral purpose, and that prefers the political status quo, business profits, and personal comforts to the traditions on which our republic was founded.

Rich and endowed as we are, the dominant concerns of our leadership have been almost wholly defensive. Our foreign policy has been dominated by sterile anti-communism and stupid wishful thinking, our domestic policy by fear of inflation and mistrust of government. We offer aid less to help others than to shield ourselves. . . . Our leaders tell us in effect that if we can just balance the budget and produce more consumer goods, the Soviet challenge will somehow disappear.

If the Eisenhower administration and the Republican politicians did not fully understand the magnitude of the Russian challenge and the nature of the world crisis, non-candidate Stevenson made it clear that he did understand both. "It is impossible," he said, "to spend years travelling around the world, as I have, without a disquieting awareness of the thrust and purpose of Soviet society. Its leaders believe in their revolution as the leaders in the American Revolution believed in theirs." The implication was plain—if the Americans would arouse themselves to reassert their revolutionary spirit in the new crisis of the modern age, Stevenson was there and willing to lead them. Or so, at least, it seemed to his overflow audience, who cheered him on, and to the seasoned reporters who were covering the occasion.

In a few days new "Stevenson for President" committees were springing up all over the country; political figures like Senators Mike Monroney and John Carroll, former Senator Herbert Lehman, former Air Secretary Thomas K. Finletter, and Mrs. Franklin D. Roosevelt were calling for his nomination; and respected columnists like Walter Lippmann were advising the Democrats to nominate Stevenson, with Kennedy as his running mate. "Stevenson," said Lippmann, "has been the successful governor of a big state, has had considerable experience in diplomacy, has had a deep indoctrination in American affairs in two grueling campaigns against an unbeatable opponent, and, all in all, is greatly respected and widely admired in all the corners of the globe."

But Stevenson characteristically did nothing further during the next month to keep himself in the political spotlight. The momentum given the draft movement by his return from Latin America and his Charlottesville speech soon slowed down for lack of more positive action by himself, and came to a grinding halt on May 10 when Senator Kennedy scored an unexpected and overwhelming victory over Humphrey in the West Virginia primary. In a strongly Protestant state, where his Catholicism was expected to count heavily against him, Kennedy won so decisively that the race for the nomination appeared to be all but over. Organization leaders were either openly declaring for Kennedy or guardedly hedging against his probable success.

The grass-roots sentiment for Stevenson, as evidenced by mushrooming local "draft" clubs, apparently still exceeded any corresponding sentiment for Kennedy, but could not be effective without political leadership. The growing group of prominent men and women, like those mentioned above, who were declaring for Stevenson, had minimal influence in state party organizations where delegates are chosen. The only party leaders of major importance who remained as possibilities to give Stevenson actual delegate strength were Governors Brown of California and Lawrence of Pennsylvania. Of these "Pat" Brown had incurred a great obligation to Kennedy by the latter's decision not to contest the California primary, leaving the way open for Brown to head a "favorite son" delegation. His high regard for Stevenson was well known, but it seemed unlikely that he would move actively against Kennedy. David Lawrence, on the other hand, was a long-time friend and devoted supporter of Stevenson. He had on a number of occasions expressed the belief that Stevenson was the best-qualified man in the party and the country, and that he should not be blamed for his two defeats by Eisenhower. But Lawrence had never specifically advocated a third nomination. Now he was on tour abroad and remained noncommittal. Under these circumstances the potential Stevenson candidacy would certainly have died— perhaps with a Stevenson endorsement of Kennedy—had not the entire situation been altered abruptly by the revelation of the U-2 spy plane disaster over the Soviet Union and the subsequent collapse of the summit conference.

The arrest of U-2 flyer Powers in the Soviet Union was announced while the West Virginia ballots were still being counted. Two days later, addressing the Conference on World Tensions at Chicago, Stevenson spoke frankly but moderately:

> In spite of all the rhetoric of the past few days, no one questions the necessity of gathering intelligence for our security. The Russians, of course, do the same, and they have a great advantage because of their addiction to secrecy, while our countries are virtually wide open to all the world's spies. But our timing, our words, our man-

agement must and will be sharply questioned. Could it serve the purpose of peace and mutual trust to send intelligence missions over the heart of the Soviet Union on the very eve of the long awaited Summit Conference? Can the President be embarrassed and national policies endangered at such a critical time by an unknown government official?

The following week Stevenson testified before Congress on his proposal that the television networks should be expected to facilitate the presidential election campaign by providing several hours of prime time for joint appearances of the candidates to debate the issues before the whole nation. The adoption of this proposal by congressional resolution and the co-operation of the networks themselves, it may be added, surely played a crucial part in the narrow victory of Senator Kennedy in November. But at the moment in May the significance of the hearing was overshadowed by the dramatic announcement of Khrushchev's wrecking of the summit. The news was brought by Senator Hugh Scott to the committee session where Stevenson was testifying before a packed gallery and a full committee. For the moment his only comment was, "This is terribly sad news." But there seemed no doubt in the minds of the crowds which gathered wherever he appeared that day in Washington, and on following days, that the sad news had returned Stevenson himself to the center of the Democratic presidential picture. Later in the day Stevenson joined his fellow leaders of the Democratic party, Speaker Sam Rayburn and Senate Majority Leader Lyndon Johnson, in a cable urging Khrushchev to reconsider, and asserting American national unity.

On Thursday, May 19, after much thought but apparently little consultation even with close friends, Stevenson made one of the most powerful political speeches of his career, and added another tag of identification to the baggage his

name will carry through the history of his time—"the crowbar and the sledgehammer." In a current issue of a popular
magazine former President Truman had once again attacked
Stevenson for being "indecisive." As it turned out, Truman's
timing could hardly have been worse if his purpose, as seems
to have been the case, was to remove Stevenson from the
path Truman's candidate, Senator Stuart Symington, was
pursuing toward the nomination. For Stevenson's speech at
the Cook County (Chicago) Democratic dinner was certainly
the most decisive of the 1960 campaign. Because it excited
so much controversy among citizens generally, and forced
presidential candidates on both sides to take up unambiguous positions, it is important to recall exactly what Stevenson
said, and it will be helpful to quote most of the central portions of this very short address:

> Premier Khrushchev wrecked this conference. Let there
> be no mistake about that. When he demanded that Presi-
> dent Eisenhower apologize and punish those responsible
> for the spy plane flight, he was in effect asking the Presi-
> dent to punish himself. This was an impossible request,
> and he knew it.
>
> But we handed Khrushchev the crowbar and the sledge-
> hammer to wreck the meeting. Without our series of blun-
> ders, Mr. Khrushchev would not have had a pretext for
> making his impossible demand and wild charges. Let there
> be no mistake about that either.
>
> We sent an espionage plane deep into the Soviet Union
> just before the summit meeting. Then we denied it. Then
> we admitted it. And when Mr. Khrushchev gave the Presi-
> dent an out by suggesting that he was not responsible for
> ordering the flight, the President proudly asserted that he
> was responsible. On top of that we intimated that such
> espionage flights over Russia would continue. (At this

point if Khrushchev did not protest he would be condoning our right to spy—and how long could he keep his job that way?) Next we evidently reconsidered and called off the espionage flights. But, to compound the incredible, we postponed the announcement that the flights were terminated—just long enough to make it seem we were yielding to pressure, but too long to prevent Mr. Khrushchev from reaching the boiling point.

And, as if that wasn't enough, on Sunday night when there was still a chance that DeGaulle and MacMillan could save the situation, we ordered a worldwide alert of our combat forces! Is it unreasonable for suspicious Russians to think such a series of mistakes could only be a deliberate effort to break up a conference we never wanted anyway?

We Democrats know how clumsy this administration can be. We are not likely to forget the fumbles that preceded the Suez crisis on the eve of the 1956 election.

But nothing, of course, can justify Mr. Khrushchev's contemptuous conduct, especially after President Eisenhower had announced that our espionage flights had been called off. But his anger was predictable, if not his violence. How would we feel if Soviet spy planes based in Cuba were flying over Cape Canaveral and Oak Ridge? And also we could predict with certainty his efforts to use the situation to split the Western Alliance and intimidate the countries where our bases are situated. . . .

It is particularly regrettable that this happened in an election year. And we can already predict what the Republicans will tell the people in the months ahead.

They will say that President Eisenhower's patience and dignity in Paris scored a diplomatic triumph by exposing Khrushchev's insincerity.

They will say that the Russians are hoping that a

"softer" Democratic President will be elected in November.

They will tell the people that a vote for the candidate the Rusisans distrust is a vote against appeasement.

It will be our duty, it will be the duty of all thoughtful, concerned citizens to help retrieve the situation and to face the hard, inescapable facts; that this administration played into Khrushchev's hands; that if Khrushchev wanted to wreck the conference our government made it possible; that the administration acutely embarrassed our allies and endangered our bases; that they have helped make successful negotiations with the Russians—negotiations that are vital to our survival—impossible so long as they are in power.

We cannot sweep this whole sorry mess under the rug in the name of national unity. We cannot and must not. Too much is at stake. Rather we must try to help the American people understand the nature of the crisis, to see how we got into this predicament, how we can get out of it, and how we can get on with the business of improving relations and mutual confidence and building a safer, saner world in the nuclear age.

The speech, inevitably, provoked a storm. Stalwart Democrats like James A. Farley equated it with treason. The conservative press unanimously, and the Republican press generally, denounced it, predictably, as a blow to national unity. A few editorial voices, like the *St. Louis Post-Dispatch,* the *Louisville Courier-Journal,* the *Milwaukee Journal,* and *New York Post,* responded with thanks to Stevenson for clearing the air and telling the truth. A few papers, like the *Post-Dispatch* and the *Denver Post,* called for his nomination; most deplored the possibility. But all agreed that the speech made him a candidate whether he meant to be or not.

The active candidates were inescapably badgered by the

newsmen, and reacted, again predictably, without relish for the situation Stevenson had created. Nixon deplored Stevenson's "damage to national unity" and expressed assurance that not only Republicans but "right-thinking Democrats" would stand with the President. Johnson thought Stevenson had gone too far, though he agreed that mistakes had been made by the administration. Symington stumbled equivocally. Only Kennedy was forthright in his agreement and his manly acknowledgment that the President ought to have made some gesture to Khrushchev before it was too late.

Into Stevenson's law office poured a mighty stream of letters from plain citizens, exhilarated or outraged. No count was made of these spontaneous letters—there were so many thousands that they had to be acknowledged by form replies, one for those who agreed and one for those who did not. And, of course, mistakes were made. "Dear Adlai," wrote one correspondent, "but I did not agree! What you need is another form letter!" What was unusual about this particular unpersuaded citizen was that he was nevertheless for the draft of Stevenson "as the Democrats' last hope for a qualified candidate"!

But if the sample of public opinion that impressed itself on Stevenson in the wake of the "crowbar and sledgehammer" speech is a fair reflection of the whole nation, as it probably was, it appears that he had once more struck an honest but painful note and failed to carry the majority with him. No accurate estimate is available, but the sheer bulk of letters criticizing his views seemed to be slightly greater than the bulk of those favoring them. Other measures of public opinion were not strictly comparable, but Gallup's polls showed a small majority supporting the President and a larger majority favoring espionage to protect the national security, even at such risks as the U-2 type of operation entailed.

If Stevenson had meant his speech as a maneuver toward

active candidacy it appears, on balance, that the effort was misguided. If, on the other hand, he intended only to take advantage of his unique position as a party leader not seeking the nomination, to say what was too risky for the candidates and what would not be said by those in power, he succeeded well indeed. In perspective it makes no difference what his calculations may have been, since he spoke, in any case, in his accustomed role as the conscience of his party and of the nation.

After Stevenson spoke, the issue had to be faced. People had to think about the facts and either explain them away, if they could, or propose ways to avoid such disasters in the future. Senator Everett Dirksen, for example, seized upon rumors that Khrushchev was hoping for the election of Stevenson to the Presidency to blame the summit collapse upon Stevenson himself! Others took up the suggestion, and much was made of an unauthorized "interview" by a French journalist purporting to show that Stevenson was ready to turn Berlin over to the Communists. At the Chicago Council on Foreign Relations, May 26, Stevenson took pleasure in correcting the record by advertising his new book, *Putting First Things First,* where, he pointed out, one could find on page 22 his oft-repeated statement that "under no circumstances will we forsake the people of free Berlin or yield to Communist threats."

Whether there was any real chance of his being nominated, indeed whether he held any hopes of a draft, Stevenson's course of conduct was probably better calculated than any other he could have devised to underscore once again his qualities of leadership, and thus to keep the question of his nomination open. The controversy he intensified over the U-2 incident provided not only others but also himself with an opportunity to offer positive proposals for foreign policy. Speaking to the convention of the Textile Workers Union on

June 1, he outlined the main points of a foreign policy which his supporters could view as a candidate's plank and which became, in considerable measure, his party's plank. There were, he suggested, five "priorities for peace":

(1) Build up a deterrent power and a limited war capability that does not depend on the budget bureaucrats.

(2) Strengthen the political and economic unity of the Western Alliance by setting up an Atlantic Council.

(3) Join with our allies in a long-range aid program to poor nations.

(4) Make it plain that general and complete disarmament under international control is an imperative for all of us.

(5) Put first things first here at home, to show the world that freedom works in meeting basic needs for schooling, research, health, housing, urban renewal, and in guaranteeing civil rights for all Americans.

The Stevenson committees, now beginning to co-ordinate their efforts to some extent through an office opened in Washington, gave this speech wide circulation.

To all the delegates, also, went appeals from the new national Stevenson committee, headed by James Doyle of Wisconsin—statements signed by prominent citizens, copies of Louis Bean's study of Stevenson's chances against Nixon, and copies of the last article written by the respected late Senator from Oregon, Richard Neuberger. In an incisive and moving piece which appeared in the February, 1960, issue of *The Progressive,* Neuberger had written:

I do not know whether Adlai Stevenson ever will become President of the United States. The path ahead of a two-time loser is pocked with perils, and 1960 is undoubtedly his last chance. Destiny often foils those who seem most prepared for destiny's climactic events. But what I do know is this—if Stevenson does not go to the White House, millions of his fellow Americans will feel they have been robbed of their opportunity to live in a time of greatness.

During the remainder of June, Stevenson, fearing, as he said, that he had been "acting too much like a candidate," made few public appearances and no substantial political addresses. His friends and supporters around the country worked hard to get signatures on petitions, distribute literature, establish an effective national organization, and persuade the Democratic National Committee—especially Chairman Paul Butler—that Stevenson should be treated like a candidate at the convention in Los Angeles, with all of a candidate's privileges and facilities. On the latter effort they were conspicuously unsuccessful, with what some thought were decisive consequences.

Stevenson supporters, like Doyle and Senator Monroney, traveled as much as they could about the country to call on political leaders, but they felt keenly the inadequacy of their efforts—for lack of manpower—as contrasted with those of the Kennedy organization. But above all they felt the lack of a candidate. More than one of the leadership group recalls that his only direct contact with the "candidate" was at some crisis or other when Stevenson appeared to be on the verge of disowning the draft movement and perhaps endorsing Kennedy.

There was never any question that Stevenson would accept if nominated. After the summit collapse, indeed, it seems probable that he would have welcomed the nomination. But he never reversed his long-held position that he should not introduce himself into the race. Even his availability seemed doubtful to a good many people, and the Stevenson campaign committee tried several times to persuade Stevenson to correct that impression. In particular, Mrs. Roosevelt undertook, through her nationally syndicated column, to present a picture of a more positively available Stevenson. But her wire to the non-candidate drew from him little more than confirmation that he was a non-candidate. He simply ignored a telegraphed appeal from Humphrey to announce himself as a candidate. By convention time his availability rested in the public mind only on his established character as no evader of responsibility and on the simple, if not overly impressive, fact that he had not disowned the efforts of his friends or of the grass-roots committees.

Despite several million signatures on petitions, newspaper advertisements sponsored by committees of scholars and educators (including many of the most eminent in the nation), by public personalities, and by plain citizens, despite active state Stevenson organizations all over the country, and despite endorsements from such influential papers as the *St. Louis Post-Dispatch*, the *New York Post* and the *Milwaukee Journal*, and liberal magazines like *The Progressive*, it was evident that Stevenson first-ballot delegate strength remained slight and that the regular party organizations were not responsive to these expressions of popular sentiment. There appeared by July to be little prospect that the amateurs could confound the professionals massed behind Senator Kennedy. What was remarkable, in the circumstances, was that the apparently and perhaps actually hopeless odds never discouraged the Stevensonians anywhere. They converged on Los Angeles with colors flying and dedicated to their cause with at least as much fervor as they had displayed in the campaign of 1952. More than one commentator has, in fact, compared 1960 to 1952, to the disparagement of the 1956 Stevenson effort.

V

While his role as candidate or non-candidate occupied the attention both of politicians friendly to him and partisans of others, Stevenson's influence was felt, anonymously perhaps, but more enduringly, in a different way—and at both conventions. To a remarkable degree the intellectual climate prevailing in the conventions of 1960 was affected by the leadership Stevenson had given during the decade then ending, by the criticisms he had made of the Eisenhower administration, and by the ideas he had advanced and the policies he had suggested and defended. The platform committees of both parties produced documents significantly more liberal, especially in the international field, than would have been likely a few years earlier. And from the planks of both parties Stevenson's voice could be heard. The Republicans, for example, recanted their "catastrophic nonsense" line of 1956 and called for suspension of nuclear testing. And in their plank on foreign aid they picked up the Stevenson proposal of 1957, which the administration had then rejected, in order to call for a program of "inviting countries with advanced economies to share with us a proportionate part of the capital and technical aid required."

But naturally enough it was the Democratic platform which more closely echoed the familiar Stevenson language, filtered to some degree through statements of the Advisory Council. Here, for example, are the opening paragraphs on "The Underdeveloped World":

> To the non-Communist nations of Asia, Africa, and Latin America: We shall create with you working partnerships, based on mutual respect and understanding.
>
> In the Jeffersonian tradition, we recognize and welcome

the irresistible momentum of the world revolution of rising expectations for a better life. We shall identify American policy with the values and objectives of this revolution.

To this end the new Democratic Administration will revamp and refocus the objectives, emphasis and allocation of our foreign assistance programs.

The proper purpose of these programs is not to buy gratitude or to recruit mercenaries, but to enable the peoples of these awakening, developing nations to make their own free choices.

As they achieve a sense of belonging, of dignity, and of justice, freedom will become meaningful for them, and therefore worth defending.

Again, on "Peace," the Democratic platform pledged:

A primary task is to develop responsible proposals that will help break the deadlock on arms control.

Such proposals should include means for ending nuclear tests under workable safeguards, cutting back nuclear weapons, reducing conventional forces, preserving outer space for peaceful purposes, preventing surprise attack, and limiting the risk of accidental war.

This requires a national peace agency for disarmament planning and research to muster the scientific ingenuity, coordination, continuity, and seriousness of purpose which are now lacking in our arms control efforts.

But if the platforms of the two political parties seemed somehow haunted by the ghostly presence of Adlai Stevenson's thought, the man himself was no ghost, as the great crowds that greeted him in Los Angeles abundantly testified. Crowds far exceeding those surrounding other candidates met him at the airport, saw him arrive at his quarters, followed him on the street, and stood outside the convention hall for days, shouting their enthusiasm for Stevenson. Southern California, the place of his birth, had always been a Stevenson stronghold, but in July, 1960, there seemed to be more extravagant feeling for him there than ever.

Stevenson had not anticipated such overwhelming receptions and was deeply moved by them. The national draft movement and the temper of the crowds in Los Angeles put upon him almost irresistible pressure to alter his position. In the end, indeed, he did in a sense acknowledge that he was a candidate. Informal overtures to him to place Senator Kennedy's name in nomination were painfully turned aside because, he felt, there were too many people too sincerely devoted to his own candidacy to warrant his turning his back upon them. Once again, and finally, he went into the whole business of his purposes, meaning, and intention. This time, on the eve of the convention's actual opening and on nationwide television, he revealed something of the feeling his supporters had aroused in him. In answer to a reporter's question he said:

> I feel a profound sense both of indebtedness and gratitude to all of these people all over the country who have worked with such unremitting diligence and effort and at such a painful price, sometimes a price not only in terms of money, but of time, and even of political jeopardy, to further this draft movement. I feel deeply, deeply grateful to them. After all, you know, when you have been in politics as long as I have, when you have been the candidate of your party twice for President and been badly defeated, and you find that there is still a residue of support and of confidence throughout the country, it can't help but touch you very deeply, and it does me.

In view of this kind of feeling a reporter suggested that he ought to "declare himself in or out, for the sake of these people." Stevenson's reply was sufficiently categorical: "I don't think it makes very much difference. They have declared me in." But the newsmen were not satisfied, and one of them put the question, for the last time, in these terms:

> Governor, is there any way that any of your supporters or any reporter on this panel could twist your arm to make you say you would be a candidate?

This was Stevenson's answer, only hours before the convention opened:

> Let me put it this way, if I can . . . that while I am personally not a candidate, and I don't believe you can persuade me to be a candidate, and I do insist on being consistent about it, I think that these supporters of mine have converted me into a candidate. That is, I am their candidate, the candidate, I dare say, of a great many people around the country who signed these petitions.

He said nothing of the warmly encouraging letters from thousands of citizens who had written to him in Chicago, urging him to run again. But the non-candidate did now at least acknowledge that he was the candidate of his supporters. And it is perhaps a diverting exercise in logic to consider which could be more accurately called a candidate, a man who had declared himself but had no supporters, or a man who refused to declare himself yet had supporters?

And he was available. "If they want me to lead them," Stevenson said, "I shall lead them."

As the convention proceedings began there seemed to be

only one possible obstacle remaining in the way of a Kennedy sweep on the first ballot—Governor Lawrence of Pennsylvania. It was not certain whether a Lawrence declaration for Stevenson would carry the major part of Pennsylvania's eighty-one votes, or even if it did, that the trend would be slowed or stopped. Neither the Symington nor the Johnson campaigns had been as successful as had been anticipated— Symington, in fact, had all but disappeared from consideration. Thus only a revision of plans by the big states, if Lawrence wished to try to lead them, could stop Kennedy and bring about the deadlock necessary for Stevenson to be nominated.

This issue was settled shortly after Governor Lawrence arrived in Los Angeles. After a long private conversation with Stevenson, Lawrence announced that he would support Senator Kennedy. With this announcement the contest seemed to be over. But the Stevenson people would not give up. Again they found themselves with the double problem of stopping Kennedy and persuading their own man not to take action on behalf of Kennedy. That the Stevenson campaign remained alive during the first three days of the convention was owing mainly to three ingredients in the picture—the deep conviction of many people that Stevenson was simply better qualified for the Presidency than anyone else, the concern in the minds of many others that Kennedy might not be quite ready for so high a responsibility, and the unprecedented enthusiasm produced by Stevenson's unexpected appearance on the second day of the convention to take his seat with the Illinois delegation. That scene, which released a torrent of emotion— much of it, evidently, in the form of tribute to a departing leader—was witnessed by millions of people on television screens and brought forth a new flood of telegrams urging delegates to vote for Stevenson.

Under the circumstances, the Stevenson managers took

courage for one final effort. The most accurate delegate count
they could make showed that Kennedy was still some fifty
votes short of a first ballot victory. If it should go to a second
ballot it was well known that Kennedy's losses might exceed
his gains, as many delegates under one ballot commitment
would move into the camps of other candidates, including
Stevenson. It was thus probable that Kennedy could not
make it unless he made it immediately, and there seemed no
serious likelihood that anyone other than Stevenson would
be able thereafter to obtain a majority. Indeed, the Kennedy
managers were frank to say that they always proceeded on the
assumption that their only real opponent was Stevenson.

In 1956 Stevenson's name had been placed in nomination
by John Kennedy. Now, as the contest between them—sought
by neither—approached its climax, Stevenson could not return the compliment. The problem, rather, was to decide who
should place Stevenson's name in nomination. Mrs. Roosevelt
and Senator Humphrey were suggested among others. In the
end, Humphrey proposed his colleague, Senator Eugene J.
McCarthy of Minnesota. It was a felicitous choice, for the
extemporaneous address McCarthy delivered to the convention was at least as moving as any since Franklin Roosevelt
called Al Smith the "happy warrior."

McCarthy's tone was that of moral exhortation, nicely
adjusted to the task of recommending to the convention that
it name as Democratic candidate for President the voice of
conscience in American politics:

> I say to you the political prophets have prophesied
> falsely in these eight years. And the high priests of govern-
> ment have ruled by that false prophecy. And the people
> seem to have loved it so.
>
> But there was one man—there was one man who did
> not prophesy falsely, let me remind you. There was one

man who said: Let's talk sense to the American people.

What did the scoffers say? The scoffers said: Nonsense. They said: Catastrophic nonsense. But we know it was the essential and the basic and the fundamental truth that he spoke to us.

McCarthy emphasized Stevenson's prophetic genius and, above all, his humility:

He said, this is a time for greatness. This is a time for greatness for America. He did not say he possessed it. He did not even say he was destined for it. He did say that the heritage of America is one of greatness.

And he described that heritage to us. And he said, the promise of America is a promise of greatness. And he said, this promise we must fulfill.

This was his call to greatness. This was the call to greatness that was issued in 1952.

He did not seek power for himself in 1952. He did not seek power in 1956.

He does not seek it for himself today.

This man knows—this man knows, as all of us do from history, that power often comes to those who seek it.

On the contrary, the whole history of democratic politics is to this end, that power is best exercised by those who are sought out by the people, by those to whom power is given by a free people.

In his peroration McCarthy, almost as though he knew he was speaking in a lost cause, implored the delegates not to turn their backs on Stevenson:

And so I say to you Democrats here assembled: Do not turn away from this man. Do not reject this man. He has fought gallantly. He has fought courageously. He has fought honorably. In 1952 in the great battle. In 1956 he fought bravely. And between those years and since, he has stood off the guerrilla attacks of his enemies and the sniping attacks of those who should have been his friends. Do not reject this man who made us all proud to be called Democrats. Do not reject this man who, his enemies said, spoke above the heads of the people—but they said it only because they didn't want the people to listen. He spoke to the people. He moved their minds and stirred their hearts, and this was what was objected to. Do not leave this prophet without honor in his own party. Do not reject this man.

I submit to you a man who is not the favorite son of any one state. I submit to you the man who is the favorite son of fifty states.

And not only of fifty states but the favorite son of every country in the world in which he is known—the favorite son in every country in which he is unknown but in which some spark even though unexpressed of desire for liberty and freedom still lives.

This favorite son I submit to you: Adlai E. Stevenson of Illinois.

There was a great demonstration—the most prolonged and enthusiastic of the convention. But the realists quickly observed that the noise and the press of people came from outsiders and from the galleries—not from delegates. Thousands of Stevenson people had managed to pack the galleries by the simple expedient of asking for tickets at the Kennedy headquarters!

But the fundamental business of conventions is not the demonstrations, nor even, perhaps, the nominating speeches. It is the balloting. This is not the place to set forth the inner history of the 1960 Democratic Convention, though such an account would contain a number of useful lessons about presidential politics. It is enough here simply to observe that despite his great lead Senator Kennedy's nomination was not the foregone conclusion it was made to appear by his managers and the news commentators. A string of "ifs" whose outcome no one could have foretold shows how close, in fact, the issue really was. The last-second decision of Alaska to go for Kennedy, with Senator Ernest Gruening invited to make a seconding speech, and the similar decision of North Dakota (both made by tiny majorities under the unit rule); the decision of Parliamentarian Clarence Cannon, releasing the delegates of Iowa and Kansas from their obligation to support favorite sons on the first ballot; and the switch during the balloting itself of certain Wyoming delegates—these apparently unrelated developments produced the margin of Kennedy's victory. Stevenson people could scarcely be censured for asking what would have happened *if*.

Thus on one ballot the leadership of the Democratic party was transferred from Stevenson to John Kennedy. Because the issue was settled by one ballot it is impossible to judge how much potential strength Stevenson actually had. The 5 per cent of the vote he did poll came in largest part from California, with only slim scatterings of votes elsewhere. But the strategy of the Stevenson leaders had been from the beginning predicated on a deadlock. Their effort had never been to try to match, or even to approach, the first-round strength of Kennedy, or of Johnson, or of Symington. As James Doyle put it, "Stevenson's supporters wanted him to get the nomination by the votes of delegates from every state and drawn from the initial supporters of every candidate." They wanted him to be the man to whom the delegates would turn for leadership if they could not otherwise agree. Nomination obtained in any other way, his managers felt, would be worthless to Stevenson and would presage his defeat for a third time. No Stevensonian wanted to see that happen.

And so there were tears but no regrets in the Stevenson camp. Meeting next day with the delegates who had voted for him and the people of the unauthorized organization that had worked for him, Stevenson spoke extemporaneous words of gratitude and appreciation, and sent them away cheering with this characteristic sentiment:

> You have given me something far more precious than the nomination; you have taught me a lesson I should have learned long ago—to take counsel always of your courage and never of your fears.

The meaning of the transition from Stevenson to Kennedy for the Stevensonians and for the nation was inevitably the subject of much press speculation and endless conversation among interested citizens. One of the Stevensonians, who preferred to be anonymous, put it—perhaps definitively—this way:

> There has been a great deal of talk here this week, after the intensity of the Stevenson demonstrations, about the emotional wrench for any political party in the "casting off of an old leader," when the ties of affection and commitment are still strong. But I don't see it as such a free act, or that the "old leader" didn't have as big a hand in the decision as the new one, admitted or not. Stevenson, and Stevenson alone, stood between the old and the new in his party; he and he alone could have barred the way, because nobody else was on that bridge. It was 1952, not 1960, that marked the end of an era, the beginning of the end for the old machines, the rising of a new order. Those who think it happened Wednesday night in the Los Angeles

Sports Arena should read a few old clippings from 1952. A new kind of "engaged" citizen, who had never participated before, rose in his wake by the thousands, many of us whose lives will never be quite the same again, just because of him. These dazzling young professionals the press has gone ga-ga over this week tell us Stevensonites that we're clinging to a dream that is past. The crazy children! They're that dream coming true! Only just not so fast, so fast, that they almost frighten their elders. This diamond hardness isn't *quite* what we had in mind.

# A Note on the Sources

THE OCCASIONAL footnotes acknowledge quotations from magazines, newspapers, or documents used only once or twice. The bulk of the material is drawn from original sources, as follows: Stevenson campaign speeches and documents, as handed to the press by his press secretary; noncampaign speeches and documents, as preserved in the Stevenson papers in Governor Stevenson's Chicago law office; Stevenson press conference statements from the transcriptions issued by his press secretary or from his law office; other material from Stevenson's books, as indicated in the text (*Major Campaign Speeches of 1952,* New York: Random House, 1953; *What I Think,* New York: Harper and Bros., 1956; *Call to Greatness,* New York: Harper and Bros., 1954; *The New America,* New York: Harper and Bros., 1957; *Friends and Enemies,* New York: Harper and Bros., 1959; *Putting First Things First,* New York: Random House, 1960); quotations from Stevenson's correspondence made available to me from his personal files.

The quotations from President Eisenhower are drawn from his press conferences as issued by the White House Press Secretary; from his campaign speeches and papers as made available to me by the Republican National Committee; from his state papers as published in the *New York Times* and in *Public Papers of the Presidents of the United States, Dwight D. Eisenhower, 1957,* Washington: U.S. Government Printing Office, 1958.

Quotations from Secretary Dulles' addresses and press conferences are drawn from the *New York Times.*

Factual information regarding historical events is regularly drawn from the reports of the *New York Times,* as indicated in the text.

The quotations from Senator Eugene J. McCarthy's 1960 speech nominating Stevenson are drawn from the tape recorded text supplied to me by Senator McCarthy.

# Index